The Young
Henry Adams

The Young Henry Adams

BY

Ernest Samuels

THE BELKNAP PRESS OF

HARVARD UNIVERSITY PRESS

Cambridge, Massachusetts

1 9 6 7

Distributed in Great Britain by
Oxford University Press, London

Printed in the United States of America

TO MARY AND ALBERT

Preface

Henry Adams failed as a politician and statesman and held up that failure, in the preface to The Education of Henry Adams, *as an object lesson to "young men, in universities and elsewhere." The precise point of that lesson continues to be obscure, for Henry, as his brother Brooks long suspected, was "never quite frank with himself or with others." In the course of* The Education *he almost succeeds in concealing the fact that that failure was a stroke of good fortune, as it freed him for a career more notable than any which gave luster to the politics of the Gilded Age.*

With stubborn humility he scoffed at his reputation as a teacher, historian, and social philosopher. In 1893 when the Loubat Prize was awarded to his History of the United States *he modestly protested that it should more properly have gone to Mahan's* Naval Warfare *or Nicolay and Hay's* Abraham Lincoln. *Literary historians have ignored his self-depreciation and have assigned him an important place in American letters. Critics following in the wake of discerning book collectors have rediscovered his two novels—*Democracy *and* Esther*—as distinguished minor achievements of the eighties. And to the anthologist,* The Education, *the* Mont-Saint-Michel and Chartres, *the* History of the United States, *and even the forgotten* Life of George Cabot *have become rich preserves of singularly moving prose. Not long ago Yvor Winters surmised that the* History *was "the greatest historical work in English, with the possible exception of the* Decline and Fall." *The praise may be* *excessive, but at least it suggests that the career of Henry Adams deserves to be studied as something other than a symbol of the pessimism of our time.*

Almost every account of his career has been colored by the ironic hindsights of The Education, *and the note of self-mockery has long discouraged the prosaic spadework out of which a sounder understanding of his accomplishment must come. Now that nearly a third of a century has gone by since his death at eighty, it seems time to explore objectively the formative period of his career from beyond the strangely blighting shadow of* The Education.

This book attempts to tell the story of the earlier Henry Adams as it may be reconstructed from contemporary materials, especially from those relating to his intellectual and literary growth. It takes him through the apprentice years which reached from the Harvard of 1854 to the Washington of 1877, years in which, after a brief revival of the political fortunes of the Adams family, a decline began, impossible to arrest and culminating in the Campaign of 1876, which proved the tomb of all their political aspirations. Thanks to these disappointments, of political hope inexorably deferred, Henry Adams was induced to follow the bent of his mind into a career of literary achievement. The record shows that despite the overpowering family drive toward public life the compulsive bias of Henry Adams's mind was artistic and intellectual. And so it happened that the careful strategy which he adopted, or, more often, was led to adopt, designed to further a political career, tended to a career progressively more remote from his ambition. He had planned to harness his ability as a writer and a scholar to his career as a statesman; in the end the statesman went to work for the historian and the novelist.

In the course of that development he not only perfected the tools of his craft—a lucid and supple prose style and the mastery of the historical method—but also sounded the main

*themes whose elaboration was to occupy the rest of his life. In
the concluding chapter I have sketched how the lines of force
traced out in this book finally attained the dynamic equilibrium
of the* History of the United States.

*I have therefore treated his early writings not only as inter-
esting in themselves, but also as valuable foreshadowings of the
writings of his maturity. I have not aimed at a definitive or complete biography of the earlier Henry Adams but have desired
rather to provide a coherent body of fact with a modicum of
interpretation which may be useful to the critical reader of
Adams's major writings.*

*The earlier and more approachable Henry Adams who tried
in vain to impose the orderliness of his mind upon the stubborn
chaos of party politics in a democracy has only recently begun
to emerge from beyond the "dust and ashes" mood of his later
life and he stands revealed as a very human and ardent idealist.
We can now see more and more clearly that the addition of the
subtitle, "An Autobiography," to* The Education, *after Adams's
death, commonly misleads the reader. His underlying purpose
was not autobiographical in any conventional sense. His own
subtitle, as he says, would have been "A Study in Twentieth
Century Multiplicity." The "manikin" of his Preface is made to
wear the Nessus shirt of a dynamic theory of history and in the
process the book ultimately broke down, succeeding neither as
autobiography nor as philosophy except in brilliant fragments.
When his brother Charles read the book in the private printing
of 1907, he good-humoredly protested the treatment of one
episode: "But you're not a bit of a Rousseau! . . . Why didn't
you let out your own most vivid recollections . . . ?" Henry
would not, for his premise was that Rousseau had disgraced
the Ego. As students of Adams know,* The Education *not only
distorts and suppresses many of the facts of his life, but it often
falls into plain factual error. For the most part therefore I have
put aside* The Education *as a primary source, referring to it*

only occasionally to indicate the sort of reservations that need to be made.

The first clear intimation of a different Henry Adams came from A Cycle of Adams Letters 1861-1865, *issued in 1920, thirteen years after the private printing of* The Education. *Subsequently the publication in 1930 of the* Letters of Henry Adams 1858-1891 *threw the door wide open upon much of his early life. Many more valuable details relating to that period have now become readily available through the publication in January 1947, of* Henry Adams and His Friends, *a collection of hitherto unpublished letters. But these collections form only a part, though a choice part, of what was an enormous correspondence. Their usefulness is also somewhat diminished by the omissions and expurgations which the editors have felt obliged to make. In an era that has seen the development of revolutionary techniques in psychological analysis, the loss of seemingly trifling details of personal and family life is distinctly regrettable.*

Fortunately a considerable amount of new material has steadily become available. Here and there in articles and biographies, in memoirs and journals, in newspapers and miscellaneous publications, other letters by and about Henry Adams and many more biographical items have come to light. Of the private papers of the earlier Henry Adams, presumably little remains, since he is authoritatively reported to have destroyed his diaries, kept since 1854, and to have recalled and destroyed many letters from his correspondents after the suicide of his wife in 1885.

A word remains to be said of the point of view adopted in this book. It has aimed at objectivity in an effort to avoid the hero worship and the mythology which is present in so much that is written about the great figures of New England. The current of American thought, as James Truslow Adams reminded us in his The Adams Family, *has inexorably diverged from the aristocratic position to which Henry Adams and his kin were committed. As a non-New Englander I have not accepted the Brahmin view of life as one that ought to prevail, however admirable or enlightened it may once have been.*

E. S.

*Northwestern University
Evanston, Illinois*

Acknowledgments

I wish gratefully to acknowledge the courtesies extended to me and the generous response to inquiries by the Massachusetts Historical Society, the Library of Congress, the State Department, the National Archives, the Harvard Archives, the libraries of Brown University, Columbia University, Harvard University, Northwestern University, Oberlin College, Temple University, University of Chicago, University of Michigan, University of Rochester, University of the State of New York, Western Reserve University, and Yale University, by the New York Public Library, the Newberry Library, the Library of the Boston Athenaeum, the Boston Public Library, the Thayer Public Library, the Thomas Crane Public Library, the Massachusetts Institute of Technology, the Historical and Philosophical Society of Ohio, the Carnegie Library of Pittsburgh, the Illinois Historical Society, the New York Times, *the New York* Post, *the New York* Sun, *the Springfield* Republican, *the* Nation, *the* New Masses, Science and Society, *and the Library Program of the Works Progress Administration, and by Professor Bliss Perry, Professor Allan Nevins, Mr. R. P. Blackmur, Mr. M. A. DeWolfe Howe, Mr. A. Henry Higginson, Mrs. Larz Anderson, Mr. Francis P. Crowninshield, Miss Elizabeth G. Norton, Mr. James D. Phillips, Mr. Elmer Davis, Mr. Tyler Dennett, Mr. George M. Elsey, Mr. Herbert R. Ferleger, Mr. Seth T. Gano, Mr. Elliott Perkins, Mr. Charles M. Storey, Mrs. James W. Wadsworth, Mr. William T. A. Fitzgerald, Register of Deeds for Suffolk County, Mr. Francis Walker, Dr. Harry A. Garfield,*

Mrs. Ward Thoron, Mr. Henry Adams, the present historian of the family, and to Mr. and Mrs. Jay S. Newcomer, who provided quiet sanctuary for the conclusion of this work.

I am under particular obligation to Professor Napier Wilt of the University of Chicago who initially encouraged the study out of which this book has grown. To Northwestern University go my thanks for the grants-in-aid which helped defray a portion of the cost of research. For her help at every stage of the making of this book I wish to express my gratitude to my wife, Jayne Newcomer Samuels.

For special permission to quote from unpublished manuscripts I am indebted to Mr. Charles Francis Adams and Mr. Francis P. Crowninshield, and for permission to quote from A Cycle of Adams Letters 1861-1865 *(1920);* Letters of Henry Adams, *edited by W. C. Ford (I, 1930; II, 1938); and* Henry Adams and His Friends, *compiled by Harold Dean Cater (1947), I am grateful to the publishers, Houghton Mifflin Company.*

CONTENTS

From the Class Life of Henry Adams. Harvard, 1858

I was born in the city of Boston, Massachusetts, on the sixteenth of February, 1838. My father, Charles Francis Adams, began life as a lawyer, but soon gave up practising and has since lived on his property. My mother's maiden name was Brooks and it was from one of her brothers that died before I was born, that I received my own name, Henry Brooks.

Any person who wishes to examine my genealogy, can easily do so by referring to the first volume of the Life and Works of John Adams, *edited by Charles Francis Adams; or by examining Drake's* Antiquarian Register, *in which there is somewhere an article tracing my descent back to some family in Wales. The subject, however, is as yet, and is likely to remain, very much open to conjecture. It is too long to introduce here.*

My mother was the youngest daughter of Peter Chardon Brooks of Boston. The genealogy of the Brooks family has also been printed, I believe . . .

Chapter One

Brahmin Pattern

1838–1858

The Boston Standard of Thought

THE CHRONICLE of Henry Adams's earliest formal schooling during the mid-forties is incongruously barren when set against the abundant record of his college days. There remains only the single vivid anecdote of *The Education* that a childhood rebellion against going to school was put down by the Jovian intervention of Grandfather John Quincy Adams. Henry knew better than to resist the man who had recently imposed his will upon Congress to force the abolition of the "Gag Rule."

School attendance was not yet compulsory under Massachusetts law, but moral obligation supplied the law's defect. In the Brahmin world the boy's duty to attend was as sacred as the man's duty to enforce attendance. Had his grandfather chosen to defend his position as he marched his grandson off to school he could have pointed out that the proper performance of this duty bore ultimately upon the fate of democracy in America. But he forebore, saving his homilies for indocile congressmen. And Henry did not encounter a fully adequate explanation of the relationship of universal education to democracy until he read de Tocqueville, and then the eulogy of the New England common school came to him with the authority of a priestly revelation. Later when he learned that de Tocqueville had twice sought out his grandfather in the early thirties to learn the New England creed at first hand, he must have felt it as a corroboration of his own stern experience. But on that summer's day in Quincy, the child Henry glimpsed no such web of relations. Sufficient that he now knew that the school of duty in which he was being bred observed no holidays. This was his first violent lesson. Like the young Cellini he would never forget the salamander in the fire.

During the long summer months he trudged down the hill from his father's house in Quincy to the country school on the far side of the village, at heart as much a Tom Sawyer as his contemporary Sam Clemens in far off Hannibal, Missouri. The "summer, as well as the immemorial family home" was in Quincy and here Henry's father built the "house on the hill" in 1833, where the family made their residence each year from "late May to early November." [1] In the winter his schoolboy's satchel and shining morning face were swallowed up each day by the schoolhouse that lay within the purlieus of the family's winter residence at 57 Mount Vernon Street in Boston. Here he recited in the bread-and-butter subjects which differed little in content from the present-day fare of the conventional grammar school.

For him education did not respect the schoolhouse threshold. It pursued him into his father's study where Charles Francis Adams, perhaps with an eye to a career of diplomacy, tutored him in French, the native tongue of his own childhood years in Saint Petersburg and Paris. Thus as a child diligently piping the accent acute and the accent grave, Henry began to feel the gravitational pull of the world's capitals.

At ten or thereabouts he enrolled in the select private school of Master David B. Tower. The school occupied a basement classroom under the Park Street Church, which lay across the Common from Henry's home. Abovestairs, in the pulpit, orthodox trinitarianism thundered its defiance to a Unitarian world which swirled past Park and Tremont streets, and that world retaliated with the derisive label "Brimstone Corner." Belowstairs, the instruction was sufficiently innocuous. Charles Francis, Jr., who had already preceded Henry on the established route to the Boston Latin School, recalled the schoolmaster as "a very portly, good-natured man," chiefly memorable as "a good teacher" commendably able to develop "the capacity for verbal memorization." [2]

From Tower's school the "family go-cart" should inevitably have carried Henry Adams to the Boston Latin School. For two centuries that public institution had prepared the natural leaders of Massachusetts—from Cotton Mather to Charles Francis Adams—and their sons, for Harvard College. It had confessedly become an academic fetish of the Adams family. But unexpectedly his father's plans for him were thwarted by an economy measure of the School Committee of Boston. "Objections were made," Henry later explained in his Harvard *Class Life*, "on account of my father's residing and paying taxes out of town, and accordingly I never joined; but was one of Mr. Dixwell's first scholars." [3]

These objections were sufficiently colorable, as the Mount Vernon Street home, though bought in 1842 for the use of Charles Francis Adams's wife by her immensely wealthy father, Peter Chardon Brooks, remained in the name of Brooks in the Suffolk County records. What his parents' rights in the house were is not wholly clear, for Bowditch in his *Records of Land Titles* tells that on the division of Grandfather Brooks's estate in 1849, the Mount Vernon real estate went to Henry's mother. In his *Autobiography* Charles Francis Adams, Jr., obscurely comments that "the purchase and occupation" of the house were "highly characteristic" of his maternal grandfather. In any event the Adamses continued to use the "gloomy, vamped up dwelling" for a winter residence until the death of Mrs. Adams in the 1880's. [4]

Master Epes Sargent Dixwell, the mainstay of the Latin school for fifteen years, suffered in the same general proscrip-

tion, as the new regulations required that the master no less
than the students should be a resident of Boston. Rather than
surrender his right to live where he pleased, Dixwell resolutely
founded a private Latin School at 20 Boylston Place. There
Henry's father sent him for a curtailed, two-year course. Benev-
olent and spectacled Master Dixwell was a lawyer who had
turned to less worldly pursuits and had prospered in them. His
interest in conchology brought him to notice as an officer of the
Boston Society of Natural History. Later, as a director of the
Society for the Diffusion of Useful Knowledge he profited from
association with such Olympian figures as Daniel Webster and
Edward Everett. Few men were better representatives of what
John Quincy Adams admired as "the Boston standard of
thought and the mental scale of Harvard." High-minded and
public-spirited, Dixwell was one of the elect who sought to
make reason and the will of God prevail. In his school young
men prepared with single-minded diligence for the Harvard
entrance examinations, studying Latin, Greek, mathematics, history, geography, English composition, and, of course, declamation. But high as the intellectual and moral aims of the school
were, the methods of instruction fell far short of the counsel of
perfection. Antiquated, stifled by a tradition of mechanical drill,
the training wearied the flesh of the Adams children and they
yearned for release. If Master Dixwell communicated any part
of his intense intellectual and scientific curiosity to young
Henry he did so because no system can wholly frustrate the inspired teacher.[5]

In the contemptuous phrase of *The Education* the years from
1848 to 1854, spent first in Tower's school and then in Dixwell's,
"he always reckoned as time thrown away." But in 1859 when
he came to compare that training with the widely admired effi-
ciency of the German Gymnasium, which he was then inspect-
ing at first hand in Berlin, he quickly recognized the superiority
of the Boston system to the stupefying regimentation of the

Germans. In Boston, teacher and pupils worked in an atmosphere of democratic easiness and freedom. There was a humane concern for individual differences. The spirit of improvement, so ably fostered by Horace Mann as secretary of the Massachusetts school system, was reflected also in the physical equipment. The Boston schools, as Henry then recalled them, were sanitary, well-aired, furnished with comfortable separate desks, and adequately supplied with globes and maps. The Berlin counterpart was unspeakably primitive.[6]

The influence of his early schoolmasters upon Henry's intellectual growth was doubtless considerable; but in those impressionable boyhood years his father's influence outweighed all else. The elder Charles Francis Adams exemplified the Boston standard of thought more perfectly than any other man of Henry's acquaintance. In his universe the paths of duty and of right were drawn like meridians of the celestial sphere and he charted his course in life and politics as if with a sextant to his eye. This moral absolutism was tempered by a Horatian acceptance of experience and tradition. He instinctively rejected the visionary projects of such Transcendental folk as Emerson, who in the crusty phrase of John Quincy Adams, did nothing more than add "some plausible rascality as an ingredient for the bubbling cauldron of religion and politics." [7] Clearly he did not own a temperament that could offer companionship to his sons for his high-minded detachment often made him an unsparing critic of their vagaries. But to Henry, at least, his remarkable balance of mind finally became an object of veneration.

In the Mount Vernon Street household the moral and intellectual influence of Charles Francis Adams visibly centered in the second floor library. This accumulation of eighteen thousand volumes, by far the largest private collection in Boston, pre-empted the finest room in the house, a sun-lit oasis in an otherwise somber dwelling. Here Henry made endless forays upon the long ranks of well-entrenched authors, ancient and modern, cultivating a love of literature that was to be a lifelong passion. Here also as a boy he first beheld the labor of the file when he patiently held copy for his father's edition of the works of his great-grandfather. With what double emphasis must the words of John Adams have entered his mind as he sat beside his father, the sounding rhetoric reviving old controversy and recalling to father and son the legacy of political responsibility which was theirs. No classroom could have made so deep an impression upon a sensitive boy as this room where the mental life of all the generations of the Adamses came to a luminous focus, a room where editorials for the Boston *Whig* were taking shape and a boy might hear the winged words of Charles Sumner, Richard Henry Dana, and John Gorham Palfrey as they debated the path of Conscience in that era of Manifest Destiny.[8]

Thoroughly imbued with lofty principles Henry Adams went out to Cambridge in July, 1854, to take the entrance examinations. For two days he busied himself with questions on the chief textbooks in which Master Dixwell had drilled him, but the ordeal was neither by fire nor water. It was quite the contrary, for one of his classmates complained that it was so easy that many worthless students passed muster.

College Justice

When Henry Adams began his studies on August 31, the course of instruction had fallen upon evil days. At any rate such noted alumni of that period as Charles W. Eliot, Phillips Brooks, and George Herbert Palmer have each claimed the distinction of having studied at Harvard during the feeblest moments of its history.[1] The damning judgment of *The Education*, however, makes the question of precise time and degree among these fellow perfectionists almost irrelevant. Yet nowhere are the sardonic exaggerations of that work more misleading than in the brilliant chapter on Harvard College.

Briefly the indictment of 1907 charges that a half century earlier "no one took Harvard College seriously," that "it taught little and that little ill"; and that "the entire work of the four years could have been easily put into the work of any four months in after life." In short, "Harvard College was a negative force." Nothing could have been farther from the fact. The four years at that maligned institution touched Adams's mind at so many points and so pervasively that even had he been a willing debtor he could hardly have listed his full obligation. For all its prescriptive character, the course introduced the student to the chief intellectual disciplines and these were taught by distinguished scholars and scientists who were in constant touch with their confreres both in America and abroad. Moreover, as the diaries of two of Adams's closest friends show, the young men did in fact take their work with astonishing seriousness.[2]

Harvard was almost as much a part of the Adams tradition as the Quincy mansion house itself. His father, his grandfather, and his great-grandfather were only the more distinguished members of a lineage which had made the mental scale of Harvard its own. One lost count of cousins and uncles and connections by marriage, unless they were as distinguished as Edward Everett and Nathaniel Frothingham or as munificent as Peter Chardon Brooks. It was Brooks, for example, who, when his son-in-law Edward Everett was president of Harvard in 1846, provided the funds for a new President's House. In the direct line, Henry's grandfather, John Quincy Adams, commanded veneration as Boylston Professor of Rhetoric and Oratory, having served from 1806 to 1809. His published lectures inaugurated the golden age of oratory in Massachusetts. There were few years, moreover, in which an Adams was not serving as a member of the Board of Overseers or as a committeeman promoting reform of the curriculum. And there was even an authentic rebel among them, Henry's long-dead uncle John, who had defied the college authorities in "The Great Rebellion" of 1823 and had forfeited his degree. In a belated act of justice the Corporation expunged the blot by awarding him the baccalaureate posthumously, a half century later.[3]

All of these associations were strengthened by nearer events. His eldest brother John Quincy, who was graduated in 1853, was a marked man from the beginning. John's first declamation in Harvard Hall, where he stood between a portrait of great-grandfather John Adams and grandfather John Quincy, moved one of his classmates to record that singularly impressive sight in his diary. Henry's second brother, Charles, now a junior, took him in as a roommate at Mrs. P. L. Storey's and insured the continuity of family admonition. Under all these circumstances Henry would have been less than human had he not felt a proprietary interest in the college or had he not sometimes looked disdainfully upon his fellow freshmen as one at least has reported.[4]

When Henry Adams came there in 1854 the college had scarcely begun to feel the first slight stirring of the wind that was one day to roar through the Yard and shake the old order of instruction to its classical roots. Change was long overdue for old reforms had become present abuses. The masterful President Josiah Quincy had introduced the Scale of Merit grading system, a system as perfect in its way as the Great Chain of Being of the eighteenth century philosophers, but successive generations of professors had become mired in its bookkeeping. Recitations proceeded as inexorably as the movement of a planet in its Newtonian orbit—and as heedless of the needs of the individual student. And the curriculum itself, in spite of a few electives, represented a closed universe of knowledge. The *vis inertiae* of the system had baffled the best intentions of two presidents, Edward Everett and Jared Sparks.

Now a new president, the Reverend James Walker, one of the most notable Unitarian preachers of the day, indicated his intention to renovate the course of instruction. His brilliant inaugural address of 1853, had it been forcefully implemented, might have stopped the mouths of critics like hard-headed Overseer George Boutwell, who proposed that Harvard be turned into a training school for "better farmers, mechanics, or merchants," [5] in which professors would be paid according to their ability to attract students. Unfortunately, President Walker lacked the ruthlessness needed to alter the system and, though Harvard was rescued from Boutwell, reform made little progress. The only significant change that Adams was to see during his four years was the introduction of written final examinations.

Outwardly the Yard lay in the timeless embrace of old custom. The most characteristic sound was the tolling of the bell atop Harvard Hall which called the students to morning prayers at seven (six, after the first Monday in April) and to evening prayers in the late afternoon. President Walker did nothing to relax the monastic rigor of this regulation until Adams's sophomore year, when gaslight was installed. Evening prayers, which had been used to keep the students from mischief, were then discontinued because they could now be given class work for an additional hour and a half. Sunday services continued unabated, with attendance strictly required both morning and afternoon. Henry's friends, recording their weary impression, if not quite the fact, would note: "Huntington preached all day," or "James Freeman Clarke preached all day." Professor Huntington, the university minister, could on occasion extort a terse "very fine" from a student and when President Walker preached in his "forcible way" the side seats were always crowded. But the unrelenting surveillance and the ceaseless warnings against whispering and fidgeting proved irksome. Adams must have resented them at least as much as his pious friend Nicholas Longworth Anderson, in whom they produced a deplorable decline of the sense of awe.

Attendance at classes and most lectures was compulsory and absences were penalized by deductions from a student's academic score. Each student's score constituted a kind of savings account, calculated according to the system of academic price-fixing called the Scale of Merit. Every recitation and every college exercise had its price; a perfect recitation earned a credit of eight points; a written exercise, twenty-four points; a sophomore theme, forty-eight points; a declamation, sixty points; a forensic, seventy-two points; and a junior or senior theme, ninety-six points. The scores accumulated from month to month, term to term, and year to year, slowly rising like an alluvial deposit, to a total, in Adams's case, of 18,580 points. The calculation worked fairly well for that part of the course which was wholly required—the work of the freshman and sophomore years. In the junior and senior years, however, the elective courses introduced disturbing variables. Once even the professor of mathematics, Benjamin Peirce, was confounded and he scrawled across a page of the grade ledger: "I do not know how to grade, and nobody can tell me."

The scale of deductions was fully set out in the copy of the Orders and Regulations which Adams received when he entered the University. If he dodged prayers wholly, he would suffer the loss of two points; if he came late, the offense, paradoxically, would cost him eight points. Lying on the grass, shouting from a window in the Yard, or playing a musical instrument during study hours might warrant a parietal admonition, penalty sixteen points. Acts of mischief rated a "private admonition," penalty thirty-two points; or, if aggravated, a "public admonition," sixty-four points. Most of the offenses fell within the jurisdiction of the detested Parietal Committee whose members served as detectives, prosecutors, and judges; and as this committee included almost half of the college faculty, the effect upon faculty-student relations was not a happy one. In this police climate the substitution of written examinations for the traditional orals seemed but one more reflection upon the manhood of the student. Bitterly, Nicholas Anderson complained, "Harvard is fast approaching to the degenerate condition of a 'high school.' " [6]

Even during the first three years of his course, when his conduct was most exemplary, Henry occasionally fell afoul of the Orders and Regulations. One episode particularly outraged Anderson. He wrote to his mother: "I am completely disgusted. Adams has been studying hard for the purpose of pulling up; he is one of the smartest men in the class . . . A few nights since he sat up late to write a theme. The consequence was he did not get up the next morning until it was too late to dress himself completely for the 7 o'clock recitation. Accordingly he went without his collar and rec'd therefor a 'private.' *This private has counteracted the study of a whole term and he now stands 34. . . .* Is college justice human justice?" In his old age Henry Adams conceded that it probably was, but by that time he was skeptical of all earthly criteria.

In the first year Adams's record shows that his behavior was faultless. At any rate he was not caught in any offense. In his second year, he strayed somewhat—deduction seventy points. In his third year, the deviation toward Avernus increased to ninety-four points, including the unjust "private"; and in the fourth year he descended vertically from grace, suffering a catastrophic deduction of six hundred and eight points. These penitential statistics place in a somewhat different light the self-disparaging statement of *The Education* that he "was graded precisely in the middle" of his class. Actually, he was no merely average scholar, like Pope's man "created half to rise, and half to fall." True he ultimately ranked forty-fourth in a class of eighty-nine on the General Scale of Merit, but his academic fortunes were extremely erratic. During his first term, for example, he was absent for an entire month, perhaps because of illness, and at the end of the term stood seventieth among ninety-one young men. He made a brilliant recovery in the second term, rising to fourteenth place in the class; but the General Scale inexorably cumulated the lean with the fat term and he placed forty-third at the end of the first year. In the second year he pulled himself up to the edge of the highest third in each term. As a result though he ended the year in thirty-sixth place, the faculty graciously included him among the thirty-three sophomores who were awarded prize books for scholarship. A renewed spurt of application in the third year raised him to twenty-first position [of eighty-nine] in the first term; but in the second term, as his interests began to shift toward Hasty Pudding theatricals, and as the parietal deductions began their fatal erosion, he fell back ten places on the term scale, losing most of his hard-won gains.

In his senior year he sought the more tangible rewards of literary competition by submitting an essay for the Bowdoin cash prize. He also improved his friendship with a half-dozen of the *bon vivants* of his class, who had long since given up hope of finishing anywhere but close to the bottom of the class and their greater sins must have comforted him. In such reckless company, Adams crossed swords again and again with the Parietal Committee; and his "somewhat noisy" ebullience which his French Instructor, Luigi Monti, had noted on the class record became open defiance.

The faculty minutes implacably record the rebel's progress. One private admonition followed another for smoking in the Yard, lounging in the Chapel, "calling up to a college window under aggravating circumstances," repeatedly cutting class in history and philosophy, and, most heinous of all, for persistently absenting himself from prayers and services. But the young heretic was not to be swerved from his apostasy; and after twenty-two more absences, he was summoned up for a public

admonition. Under the College Rules a suitably disapproving
letter was sent to his father. The episode was hardly calculated
to strengthen his position at home, where his streak of waywardness already caused some concern. Only the providential
ending of his college course saved him from the penalty of suspension or dismissal for continued absence from prayers.

His achievement in individual courses exhibited a number of
curious variations. Excluding the extreme variation of his freshman year, he averaged well within the highest third of his class
in Latin and Greek. His French profited from the early instruction which his father gave him, for he stood within the highest
quarter of the class. In German he fared less happily, falling
from the highest to the lowest third in one year. He excelled in
botany, and achieved the rank list in astronomy, but fell short
of the mark in chemistry and physics. As for mathematics, he
averaged only a little above the middle of the class. His strongest subjects were literary composition and elocution. Professor
Francis Child recognized the excellence of his compositions by
ranking him the fifth scholar among ninety-four in one course
and the second scholar in another in which his attainments as a
speaker were included in the calculation. Such recognition was
hardly the "hesitating approval" acknowledged by *The Education.*

Adams had small appetite for formal rhetoric even when it
was taught from Bishop Whately's masterly text and his defi-
ciencies in that respect offset somewhat the high scores on his
themes. In history, his extremely favorable home environment
gave him apparently no special advantages, as even in the remarkable second term of his first year he barely made the top
third in the class in Greek history. And in the reckless last year,
constitutional history went down indistinguishably in the common ruin of philosophy, ethics, logic, and political economy.
Professor Torrey, who was one day to be his colleague in the
Department of History, ranked him sixtieth; and in the course

in ethics President Walker placed the young immoralist sixty-second among eighty-seven.

In the face of such fluctuating fortunes the proud and earnest young man could not be blamed if sometimes in moments of depression he felt that "the whole thing didn't pay." After all the striving for distinction during his first three years, he had seen the prize elude his grasp and his most conscientious efforts nullified by narrow-minded disciplinarians. Happily, his later work as a scholar was to show how completely lacking in prediction value was the soulless arithmetic of the Scale of Merit. A mere dozen years later Eliot ignored his undergraduate record and invited him to join the faculty. Not to be outdone in perspicuity the Phi Beta Kappa Society made him an honorary member as, after a similar delay, it had honored his father back in 1840.[7] One trifling value did accrue from the mechanisms of the Scale; it at least gave him a useful metaphor when he came to covet a place on the "rank list" of American society.

The Mental Scale of Harvard

Looking backward on Adams's development as a thinker across the half century which separates *The Education* from Harvard College, one is struck by the persistence of certain fundamental ideas. Of course *The Education* acknowledges no such continuity. Harvard College "taught little, and that little ill, but it left the mind open, free from bias, ignorant of facts, but docile." Studied in detail, however, the intellectual climate of 1854-1858 assumes a significance in the genealogy of his thought quite different from that assigned by him in retrospect. Granted that the filiation of ideas can hardly be traced to particular passages in a textbook or ideas voiced in a lecture, textbooks and lectures do have value as soundings in the stream of thought which swept through his retentive mind. They give clues toward solving the enigmas of his later thinking and help account for its direction.

There are a number of significant—and unintended—implications in the scornful bill of particulars filed by *The Education*. "Beyond two or three Greek plays," reads the indictment, "the student got nothing from the ancient languages. Beyond some incoherent theories of free-trade and protection, he got little from Political Economy. He could not afterwards remember to have heard the name of Karl Marx mentioned or the title of 'Capital.' He was equally ignorant of Auguste Comte. These were the two writers of his time who most influenced its thought. The bit of practical teaching he afterwards reviewed with most curiosity was the course in chemistry, which taught him a number of theories that befogged his mind for a lifetime. The only teaching that appealed to his imagination was a course of lectures by Louis Agassiz on the Glacial Period and Palaeontology, which had more influence on his curiosity than the rest of the college instruction altogether."

If we place his praise of Agassiz beside his depreciation of the course in chemistry we are put on the track of his underlying philosophical attitude. Agassiz did indeed stimulate his curiosity, but toward geology and related sciences and unfortunately for the validity of his future study of scientific thought, away from the experimental and biological sciences. Agassiz maintained toward science the anti-rationalist point of view of a devoutly religious Christian and in the conflict between science and religion he bent all his learning toward supporting the theological position expounded by his colleagues President Walker in the courses in religion and Professor Francis Bowen in the course of philosophy.

The point is crucial because though religion disappeared from his early life, as he later noted, it was not replaced by a philosophy of naturalism or scientific materialism. He continued to adhere almost desperately to a philosophy of quasi-idealism. As a youth, he believed, as he said, in a "mild deism," and the will to believe remained sufficiently strong in him that he tried to "recover" in later life what he called "the religious instinct." But having become "the freest of free thinkers" he thought he had failed. However, the little known poems, "Prayers to the Virgin and the Dynamo," and the mystical aestheticism of the *Mont-Saint-Michel and Chartres* show that he succeeded in recapturing the religious instinct even if he did not recapture a tolerable theology to which to attach it. In his final bewilderment and contempt of the world and the intellect he approached the otherworldliness of the Church.[1]

In the face of the inescapable religious influences of the fifties it is understandable that a mild deist should have yielded at least a little. More than mere churchgoing was involved; that he had been accustomed to. The attitude was all-pervasive and at bottom anti-intellectual; it permeated the textbooks and lectures, and was faithfully echoed in the undergraduate magazine. In "The Limits of Knowledge" a student gravely pointed out the insurmountable physical barriers to scientific research which should give the scientific enthusiasts pause. Science had been misused to undermine religion and God and had given birth "to the varied forms of materialism and infidelity." There was "a right way to harmonize nature and religion": let science keep its proper sphere. In season and out Agassiz lectured his students on the "folly of the development theory" of Lamarck and swept away all opposition with dramatic eloquence. In his great work *Études sur les glaciers* his religious bias had led him to embrace the theory of catastrophism in geology and to oppose the materialist explanations of the uniformitarian theory advanced by Charles Lyell. He impatiently rejected the views of the evolutionists as contrary to manifest religious truth: "There seems to me to be a repulsive poverty in this material explanation, that is contradicted by the intellectual grandeur of the universe; the resources of the Deity cannot be so meagre that, in order to create a human being endowed with reason, he must change a monkey into a man." Professedly a scientist he nonetheless grasped at the acknowl-

edged imperfections in the geological record and magnified
their importance in order to discomfit the evolutionists, a device
adopted by Adams when he later reviewed the work of the Darwinist Sir Charles Lyell. Throughout Agassiz's texts one un-
ceasing burden runs: To study the data of science is to become
acquainted "with the ideas of God himself." [2]

Agassiz's intellectual dictatorship did not go entirely unchal-
lenged. Much patient scientific research and study went on in
the laboratories of Harvard, although no one was Agassiz's
equal in dogmatic argument. Professor Josiah Cooke, whose instruction in chemistry "befogged" Henry's mind, was himself
a scientist of considerable reputation, as was also Professor
Joseph Lovering, in physics. The texts which they used and in
which Henry dutifully recited were not only modern but the
work of authorities in each field: in chemistry, Dr. Julius Stöckhardt's recently translated *Principles of Chemistry* and the
even more recent *Elements of Chemistry* by Thomas Graham,
one of England's most distinguished research chemists; in physics, the *Natural Philosophy* of another English scientist, Dr.
Golding Bird; and in astronomy the most recent edition of the
Outlines of Astronomy by the famous Sir John F. W. Herschel.
True enough, authors like Dr. Stöckhardt made reverent allusions to the Almighty who "appointed the chemical processes
for his servants," but these obeisances were little more than
decency required. The overwhelming emphasis of these writers
was placed upon experimental method and they cautiously
avoided the appearance of dogmatism. "We have no absolute
knowledge of the forces of nature, they having as it were a spiritual existence" constituted Stöckhardt's preface to his sketchy
exposition of the "laws of chemical combinations."

The qualities of mind that distinguished the scientists of
Harvard are strikingly revealed in the *Proceedings* of the American Academy of Arts and Sciences. The Academy was largely
a Harvard affair and during Henry's undergraduate days the

names of Josiah Cooke, Joseph Lovering, the anatomist Jeffries Wyman, Louis Agassiz, and that "mildest of men" the botanist Asa Gray figure largely in the *Proceedings*. There were frequent reports on original research, supplemented by the reading of communications from European scientists and scholars. Into the library of the society there streamed scores of scientific publications from all quarters of the globe. Occasionally laymen of scholarly interest were elected to the Academy as Fellows and often played host to a meeting of the society. One of these distinguished amateurs was Henry's father, who was elected on February 10, 1857.

A meeting held at the Adams home on April 13, 1858, has special significance for it marked the beginning of the violent controversy between Agassiz and Asa Gray over the question of the distribution of species, a controversy that grew in bitterness and soon pre-empted the whole time of many subsequent meetings. Quite likely young Henry Adams, who was then a senior at Harvard, was allowed an inconspicuous chair at the meeting. His sympathies, we may safely conjecture, were with Agassiz, whose lecture on the Florida Reefs he had heard that morning, and with Professor Francis Bowen and the other conservatives. Gray, adopting the hint of Charles Darwin with whom he had long corresponded, argued that biological species had been disseminated from a common center of development. Incidentally, Darwin had sent him an outline of the theory of evolution on September 5, 1857. Agassiz naturally maintained the theory of simultaneous "creation" of species, a theory far less offensive to conventional religious beliefs than the one asserted by Gray. Gray was distinctly in the minority, and though colleagues like Jeffries Wyman were sympathetic they prudently avoided the anathema of Agassiz and the respectable majority.[3] Agassiz's attitude in these debates is suggested with painful clarity by a comment which he inserted in the third volume of his *Contributions to the Natural History of the*

United States. The arguments of Darwin he said "have not made the slightest impression on my mind, nor modified in any way the views I have already propounded." As for the asserted fairness of Darwin in not suppressing evidence even when it seemed adverse, Agassiz regarded that simply as evidence of Darwin's stupidity in being unable to see facts fatal to his argument.

Gray, who became Darwin's foremost American expositor, was as far from being a militant atheist as was Darwin himself. In fact, in his effort to secure the acceptance of Darwinism he initiated the classical argument to reconcile the Darwinian theory with religion. Agassiz, he protested, insists on dealing with the "why" of phenomena; Darwin simply asserts the "how." Natural selection was not inconsistent with a theistic view of the universe. That may have been true, but for all Gray's circumspection, conservative Christians did not overlook the threat to their cosmogony. Gray shrewdly appraised Agassiz's limitations as those of "an idealizing philosopher" full of "sharp and absolute distinctions." Unerringly he touched the weakness of men of Agassiz's temperament. "Most people," he said, "and some philosophers, refuse to hold questions in abeyance, however incompetent they may be to decide them. And curiously enough, the more difficult, recondite, and perplexing, the questions or hypotheses are—such for instance, as those about organic Nature—the more impatient they are of suspense." [4]

If then it was Agassiz who most stimulated his imagination and had most influence on his curiosity, we can understand why Adams profited little from the point of view presented in the physical sciences, and finally attributed his own disinclination to keep abreast of those sciences to the limitations of his instruction. Josiah Cooke, like Gray, and like Graham, the author of the *Elements of Chemistry*, knew that science progressed by an infinite number of increments. This undramatic snail's pace and excessively cautious theorizing did not attract young Adams; but Agassiz's bold rhetoric did. Even though Agassiz's piety was probably unacceptable to him, what was distinctly congenial, as the event shows, was the romantic idealism of the master, his impatience with intractable facts, his pursuit of ulterior motives.

Later Sir Charles Lyell tried to counteract this influence by giving Adams's reading of the *Origin of Species* the benefit of a Darwinist commentary, but the initial impress could not be obliterated. By a strange trick of memory Adams came to think of himself as a "Darwinist before the letter: a predestined follower of the tide" yet the name of the one American to whom he should have been drawn in such case, Asa Gray, he let slip away into oblivion. As for Darwinists Tyndall and Huxley, these survived in recollection as mere "triflers"; Agassiz however remained the hero of the piece.

It would be inaccurate to attribute to Agassiz complete credit for strengthening the ancestral bias of mind which Adams brought to Harvard. He had able coadjutors in President Walker and Professor Bowen. A distinguished representative of liberal Unitarianism, President Walker, like Agassiz, denounced the "justly called 'dirt-philosophy' of materialism and fatalism," as Francis Bowen phrased it. As early as 1853 he lectured to his students on "The Origin of Man" in order to deny "the unity of their source." His main classroom weapon against the revolutionary secularism of 1848 was William Paley's *Natural Theology: or, Evidences of the Existence and Attributes of the Deity*, the *Summa* of eighteenth century England. In the freshman course the popular president quizzed Adams and his classmates on the pseudo-rationalistic doctrines of the text, kindly helping them over the hard places when they got stuck, and "philosophizing himself" at every opportunity.[5] Paley ransacked the physical sciences for his system of supporting analogies, filling his work with the language of comparative anatomy, astronomy, and physics. For example, one of his most popular analogies declared that there was a *"vis inertiae* which keeps things in their places . . . in the civil world as well as in the material." Such use of science had of course long been a commonplace in the Adams household and the works of Paley were old standbys. Under such discipline science could not be an end in itself; its ulterior purpose must be made to serve God, the soul, or finally abstract mind.

Professor Bowen's course in philosophy radiated equally theological and anti-scientific influences. It frankly aimed to combat "the licentious and infidel speculations which are pouring in upon us like a flood." As Alford Professor of Natural Religion, Moral Philosophy, and Civil Polity, it was Bowen's duty "to demonstrate the existence of a Deity or First Cause" and this he did with the help of his recently published text *The Principles of Metaphysical and Ethical Science Applied to the Evidences of Religion.* "The work," said Bowen in his preface, "may be regarded as an imperfect supplement to the invaluable treatises of Dr. Butler and Dr. Paley, the principal object being to consider those objections and difficulties in the way of the believer which are of recent origin, or have grown out of the recent discoveries and speculations in science and philosophy." Though it concluded with such parochial topics as "The Proof of Revelation," it did introduce Henry Adams to the leading problems of metaphysics, and perhaps counteracted the effect upon him of his father's distaste for the subject.

Bowen's lectures were as persuasive as his texts and quickly won over the intellectual young men like Nick Anderson, who imparted to his mother that "metaphysics is my favorite study." There was indeed ample matter for debate among the serious-minded and argument-loving young men in their frequent gatherings at one another's rooms. The difference between physical science and metaphysics, the nature of causation, fatalism and free will, the argument from design, the nature of instinct, the goodness of God, the origin of evil, and the unity of God, all of these were topics of vital concern in their environment.

On the side of ethics, President Walker used Jouffroy's *Introduction to Ethics* to instruct the seniors in "the facts of man's moral nature" and to warn them against "the philosophical systems which, in their principles, are destructive of ethical science." He placed under the ban of his disapproval the systems of Hobbes, Spinoza, Hume, Bentham, Smith, and others, a disapproval which Adams remembered a few years later when he began to study those philosophers. Jouffroy exempted only Price and the philosophers of the Scottish School, for they sought "the rule of human conduct where truly it is to be found —in the conceptions of reason"; and they deduced "that good and evil are immutable," arising "not from sensation but from intuitive reason."

If Adams required a rationale for the school of morals in which he was bred, the school that he always believed was native to New England and Massachusetts, Jouffroy provided it. Certainly Adams never doubted that he belonged to that élite for whom Jouffroy wrote: "Those who live in the future, and who are seeking, from government and the laws . . . a new system of faith on the grand questions which must forever interest man . . ." In the realm of practical affairs the philosopher gave reassuring support to the cautious attitude which Henry and his brother Charles had begun to take toward the abolition of slavery in the South. In speaking of revolutionary movements Jouffroy said: "When we once comprehend what is really to be accomplished, we see that it cannot be done in a moment . . . revolution must be gradual." [6]

In political economy, which was taught by the versatile Bowen as "one of the Moral sciences," the doctrine was conservative and nationalistic as well. His text, *The Principles of Political Economy Applied to the Condition, the Resources, and the Institutions of the American People,* published in 1856, pro-claimed a resolutely American system, superior in almost every respect to the "aristocratic polity" of England. As an Adams, Henry no doubt agreed with the opinion that "the true causes of national decay and national prosperity" were to be found "in the character of the people and of the institutions which they live under, and not"—as was contended by Adam Smith and his English followers—"in any imaginary or real advantages or drawbacks of territory, soil, and climate." The causes of the admirable increase of capital in the United States were "moral rather than physical," the moral excellence of the American consisting in his "disposition to toil, to dare, and to save." As proof of this proposition, Bowen cited the current prosperity of Massachusetts. Unfortunately for the force of his evidence, the Panic of 1857 broke just before Adams and his classmates be-gan the study of this text.[7]

For his authority on currency matters Bowen relied chiefly on Horne Tooke, "the able advocate in England of what is called 'the banking principle' of circulation." This exposition of sound money principles supplied a theoretical basis for what Adams had long been learning in the household of his father, who was then President of the Mount Wollaston Bank and an ardent advocate of a carefully regulated currency redeemable in specie on demand;[8] it suggested to him as well a source of materials for his later studies in British and American finance.

Bowen's text presented what we may take to be the official position of the Harvard hierarchy and its conservative grad-uates. Henry's father would surely approve the principle that "statesmen have been obliged to make the study of politics sec-ond to that of political economy." And Henry, no more than his father, would have seen no reason to question the explanation of the revolutions of 1848 as "the disastrous consequences of the insane attempts" of Socialism and Communism to reorganize society. It was the object of "the great truths" of political economy to show the fallacies of equalitarianism or economic democracy; for "the only equality of condition which human nature renders possible, is an equality of destitution and suffering." There was of course much more of the same sort of mystical dogmatizing, the final import being that economic law was the highest law of the land and of the world, immutable and eternal, like the human nature on which it was founded.

It is appropriate in this connection, to recur to Adams's complaint that he did not hear the name of Karl Marx nor the title of *Capital* at Harvard in 1858. Once more memory played a self-protective trick by imputing his own default to his alma mater. It was true that the Communist Manifesto appeared in 1848 but the contents of that document were largely ignored by the commercial press and knowledge of it was confined to a few obscure German émigrés and their labor union associates. As for Marx's articles on the Eastern Question printed by the New York *Tribune* while Adams was an undergraduate, these hardly prefigured the *Capital*. Even that great preliminary study *Zur Kritik der politischen Ökonomie* was not issued until 1859 when Adams himself was in Germany. There is no evidence that either he or his companions used that work to improve their reading knowledge of German. Actually not even the first volume of *Das Kapital* was put into print until 1867, and it was not published in an English translation until 1877. Adams, himself, did not feel impelled to read the book until early in the 1890's. Interestingly enough in 1879 Marx's writings became the subject of a seminar essay at Johns Hopkins University, an essay entitled "Karl Marx, the Internationalists, and the Commune of Paris." They had undoubtedly been studied earlier in working class circles, as the Communist Club of New York was founded in 1857 and became a section of the International Working Men's Association in 1867-1869.

Adams's twin complaint that he did not hear mention of the French positivist Auguste Comte seems equally unjustified, though on quite different grounds. In the summer of 1854 when Adams was preparing to enter Harvard, Professor Bowen wrote a sarcastic review of Harriet Martineau's translation of Comte's *Cours de Philosophie Positive*. The article appeared in the *North American Review,* probably the most carefully read magazine in the Adams household and long a part of the family tradition. If somehow he missed that article, he could hardly have escaped hearing some condemnation of "the most eminent infidel philosopher of the present day" from the lips of Bowen himself. In a long footnote in the *Metaphysics and Ethics* Bowen went out of his way to denounce "a shallow and impious remark" of the great positivist. A rather different opinion of Comte was easily available to young Adams among his own books, for in October, 1857, he bought a copy of George Henry Lewes's *Biographical History of Philosophy* and the underscorings indicate that he read at least the first part with considerable care. In the preface Lewes urged "every one who takes an interest in philosophy" to read Comte's book, the "opus magnum of our age." It seems likely therefore that Adams had some knowledge of that revolutionary thinker before he read Mill's popularization of Comte in 1866.

The Department of History like the Department of Philosophy embraced a number of related subjects, such as political science, the philosophy of history, and the constitutional history of Europe, England, and America. Within this broad area, Adams supplemented his desultory reading of the eighteenth century historians on his father's shelves with the careful study of such works as Thomas Arnold's Introductory *Lectures on Modern History* and William Robertson's notable introduction to his *History of the Reign of Charles V.* Arnold defined history for the future professor of medieval institutions as "the biography of a political society or commonwealth," whose external existence is exhibited chiefly in its conflicts with other nations and whose internal life is reflected in its institutions and laws.

In his recognition of the value of institutional history and in his analysis of national character Arnold, anticipating the German influence on Adams, provided an effective antidote to the hero worship of Carlyle, for which he substituted the study of forces and movements. Slight as the impetus may have been it turned Adams in the right direction. When he came to write history, he scrupulously avoided the delusive attraction of the hero, visualizing American history "as a subject for the man of science rather than for dramatists or poets." Equally important to the potential historians in Professor Torrey's class was Arnold's emphasis on the supreme value of contemporary records and his praise of the investigative method of Niebuhr.[10]

In the time-stained pages of Robertson's elaborate *View of the Progress of Society in Europe from the Subversion of the Roman Empire to the Beginning of the Sixteenth Century,* Adams made his first important contact with that period of history in which, not more than a dozen years later, he was to become a specialist. In addition, coming earlier in his reading as it did, Robertson's brilliant synthesis of an entire age gave a more memorable model than Gibbon's introductory chapters of the *Decline and Fall.* Robertson adopted Hume's theory of the cyclical movement of history in order to present the reign of Charles V as an important turning point in European history.[11] The success of the analysis was another resource to be treasured up for the first six chapters of Adams's own *History.*

Probably the most influential text of Professor Torrey's course was Guizot's *History of the Origin of Representative Government in Europe.* Guizot's great generalizations were of a sort to delight the philosophical statesmen of the Adams school. He analyzed the rise of representative government not as a series of isolated phenomena, but as part of a great organic and progressive movement, a movement demonstrating that "unity and consecutiveness are not lacking in the moral world, as they are not in the physical." Later Adams would encounter other writers, Buckle for instance, who also sought for "a bond which may unite and harmonize" the great mass of seemingly unrelated facts of human experience and give them meaning and direction.[12]

Professor Bowen, in his discussion of "The Unity of God," came closer to a description of a science of history. He asserted that science had steadily progressed in reducing the apparent chaos of "purely physical occurrences" to laws, such as the law of chemical affinity or the law of gravitation. By analogy, "the study of history and of the laws of the human mind, with a knowledge of the fundamental principles of politics and political economy, brings order into this chaos [of "human affairs"], and makes the past intelligible, and the future a subject of calculation and foresight." [13] But here in Guizot the note was first struck; here was first indicated the nature of that Tannhauser quest upon which the prospective historian must embark. He must establish the continuity of history. Facts as such could not engage the interest of the true historian; their drift was all.

Guizot ably set forth political principles which harmonized with the creed of the Adams family. Representative government did not sanction the tyranny of a mere numerical majority; it derived its authority from reason and justice as exhibited in appropriate political capacity. Only with tongue in cheek could one say that the voice of the people is the voice of God. Hence the majority must be subjected to such checks as "electoral representation," "public legislative debate" and the free expression of public opinion. Absolute power was the root of political evil, but Guizot conceded the necessity of adequate power safely apportioned and balanced among the branches of government. Like Jouffroy, Guizot also recognized the need of a political élite: "There will always arise and exist," he said, "a certain number of great individual superiorities who will seek an analogous place in government to that which they occupy in society." So far Henry Adams's family had helped to meet that need with the highest distinction.[14]

In the realm of first principles, in science, theology, philosophy, economics, and history, these then were the chief influences that played upon the mind of Henry Adams; these were the winds of doctrine that eddied across Harvard Yard and made its intellectual atmosphere. And there is every reason to believe that young Adams willingly sailed before the wind. If in science there were appreciable counter-currents, the prevailing pressure was, so to speak, from the northeast, sternly conservative in all the "moral sciences," deeply Christian in its ethics, idealist in the drift of its philosophy, and contemptuous of alien philosophies. Adams may have felt that Harvard imposed no bias upon the minds of its students. In a sense it did not need to; it had only to erect an adequate rationale for that already there. The bias was so uniformly diffused that one might oscillate all the way from Cotton Whig to Conscience Whig and still conform. Only on rare occasions did Adams ruefully glimpse the truth, that he never truly escaped from either Harvard or Boston.

One more professor deeply influenced Henry Adams: James Russell Lowell. It was he whose graciously informal manner of teaching set a model for future imitation. Lowell, about whom clung the aura of literary success, had come to Harvard in 1856 to succeed Longfellow in the Smith Professorship of Modern Literature. To his admiring students the pithy opinions of Hosea Biglow already had the force of proverbs. His *Fable for Critics* displayed an impeccably Cantabrigian taste as his *Vision of Sir Launfal* asserted an equally irreproachable idealism. His "Ode to France" had betrayed some sympathy for 1848 principles, but that errant mood had passed. Moreover, his sentimental resentment against the erosive action of the new science upon religious belief chimed in perfectly with the opinions of his colleague Bowen.[15] Now the study of Dante and the medieval tradition was preparing him for his reverent journey to Chartres and for the confused confessions of "The Cathedral."

Fresh from a year of hard study in Europe, Lowell communicated his humanistic enthusiasms to Adams in the cosy literary seminars that met in his study and he stimulated in his protégé a desire to go to the fount of German scholarship. His course of lectures carried no college credit but, like Agassiz's, they were exceedingly popular. He reworked the successful series of lectures, given two years before at the Lowell Institute, on Chaucer, Spenser, Milton, Butler, Pope, and Wordsworth. We may guess that it was his depreciations of Wordsworth that deterred Adams from reading the poet until many years later. Lowell also supplemented the work of the Italian instructor, Luigi Monti, by quizzing his select group of humanists on Dante's *Divine Comedy*.

The brilliantly discursive criticism which Lowell heaped to the brim with allusion and satire must have shown Adams the interesting possibilities of sheer opinionativeness. Here was a critic who took few intellectual risks among the little-known writers, but on familiar ground asserted himself like an urbane sea-rover, making the law on his own quarter-deck. It was criticism that relied on the sharp and colorful epithet and solicited the quick rewards of iconoclasm. The example obviously captivated Adams, but it proved in the end to be sterile instruction. Shrewd insights and an abundant wit were scarcely the ingredients of a coherent critical system. Adams learned no other that would serve him when he became a professional reviewer himself.

Of Men and Books

The college education which *The Education* would have scornfully crammed into four months embraced a Baconian breadth of interest. Some of that range we have already examined. How much more was set before him can be quickly summarized. Greek and Latin naturally dominated the curriculum for the three years in which they were required. In Professor "Corny" Felton's class in Greek young Adams read Homer's *Odyssey*, the *Alcestis* of Euripides, the *Panegyrics* of Isocrates, and the *Ajax* and *Oedipus Tyrannus* of Sophocles. Tutor Chase took him through Livy, Cicero's *Tusculan Disputations*, and the *Odes* of Horace. It was Chase who, less indulgent than his colleagues, forbade his students to keep their books open during recitation. Under Professor Lane Adams further explored the legacy of Rome in Cicero's *Brutus*, the *Satires* of Horace, the *Annals* of Tacitus, and the *Institutes* of Quintillian. The Department of History placed these texts in their proper perspective with the study of works like Sir William Smith's recently published *History of Greece*. In mathematics, Tutor James Peirce drilled him in plain and solid geometry, trigonometry, and algebra. Not being especially proficient in mathematics, Adams was obliged to study analytical geometry with the so-called "Jackson guards" while the more proficient upper third of his class, scornfully known as "Peirce's Reliques," were admitted to the mysteries of "Curves and Functions." In the Department of Rhetoric Professor Child devoted six semesters to such works as Vernon's *Anglo-Saxon Guide*, Latham's *English Grammar*, Murdoch and Russell's *Orthophony*, Russell's *American Elocutionist*, and Bishop Richard Whately's great treatises on rhetoric and logic, all supplemented of course by a generous program of declamations, themes, and written forensics. In the field of constitutional history, Professor Henry W. Torrey relied chiefly upon Furman Sheppard's elementary *Constitutional Textbook*," "A practical and familiar exposition of the Constitution of the United States," and Bishop Whately's *Easy Lessons on the British Constitution*. Among the modern languages, he studied French, German, and

Italian, the last two being electives chosen from the small group of such courses which also included Spanish and advanced courses in mathematics and the classics.[1]

After the freshman year the daily fare of set recitations was enriched with semi-weekly and weekly lectures. Professor Felton's series on Greek literature and Professor Lane's on Latin literature were obligatory; so too were Professor Lovering's three series on statical electricity, mechanics, acoustics and optics; Professor Cooke's on chemistry; Professor Bowen's on political economy, and Professor Torrey's on constitutional history. In Adams's lost diary there were doubtless as appreciative entries as those of his chum Crowninshield: "Felton gave us a very interesting lecture about the Sanscrit Epic poems and Homer's works," or "A capital lecture on electric light with some beautiful experiments from Prof Lovering." Asa Gray gave the series on botany; Jeffries Wyman, on anatomy; Agassiz, on geology and zoology; and Lowell, on modern literature. Each man had his appreciative following.

A corrective note remains to be added to Adams's later scorn for the meager residue of the "ancient languages." Not only did this opinion overlook the resource of allusions and similes which he delighted to use like Attic salt to season his writings, it also overlooked his early pleasure in Latin poetry. In the dark days of the Civil War, for example, he sent his treasured pocket Horace to his brother to drive away the tedium of the barracks. His earlier and more favorable opinion of the value of this study appeared in his Class Day Oration of 1858. As Class Day orator he remarked that "ten years after graduating a great part of every class . . . recollect no more of Horace and Livy and Cicero than they do of Descartes and Spinoza, and care as little for the grace of the first as for the logic of the last . . . [but] . . . contact with those old pagan writers has tinged our minds with the richness of their mental dyes." Undoubtedly he shared his friend Anderson's feeling that "the older I grow, the more thankful will I be for the little learning I possess" of the classics.

The track of Adams's general reading during these years is lost in an immense maze of literary acquaintances. He accumulated a small library of his own, numbering something over a hundred titles. These he meticulously catalogued during the lazy August days in Quincy while marking time before his departure for Europe in the fall of 1858. Many of his favorite authors do not appear and for these he must have relied upon his father's collection. Washington Irving, whom he professed to know by heart, is absent; likewise Horace Walpole, whose name figures so familiarly in his early correspondence.[2]

Here and there in the books themselves many passages are neatly marked for future reference. There are none of the pugnacious marginalia, however, which were to characterize his later reading. A few of the titles have more than ordinary interest. For example, the *Essays de Henry Adams,* proves to be a handsomely bound volume of off-prints and manuscripts of his college articles. The ten volume *Life and Works of John Adams* probably came to him as a reward for his proofreading. A specially bound set of five volumes contains the political pamphlets and orations of John Quincy Adams, presumably a collection bequeathed to Henry's father. Another work from the same pen, *The Letters on Silesia,* may well have supplied the pattern for the grandson's letters to the Boston *Courier.* A nine volume set of Edmund Burke, another hereditary favorite, indicates the source of many verbal echoes in his early style. His early-and-late idol, Carlyle, was represented by *Heroes and Hero Worship* (purchased March 8, 1855), four volumes of essays (purchased in October, 1857), *The French Revolution, Past and Present,* and *Sartor Resartus.* Some of these works show many thoughtful pencillings; however, the unmarked pages of *Sartor* made a far more lasting impression upon him, as *The Education* illustrates. He did not possess the Italian version of Dante which he studied with Lowell and Monti; but he did have John Carlyle's translation, a work used by some of his less scrupulous classmates as a pony.

Among the English poets he owned editions of Dryden, Gray, Cowper, Collins, Byron, Hunt, Keats, Tennyson, and the Brownings. The novels show an equally helter-skelter choice, and are equally unrepresentative of the wide range of his reading: Smollet, Sterne, Goldsmith, Thackerary, Miss Ferrier, and George R. P. James. In American literature the representation is even more scanty and misleading, only Whittier, Longfellow, Lowell, Ik Marvel (Donald G. Mitchell), and Prescott.

Other indications of his reading or, at any rate, of his literary taste occur in a *Harvard Magazine* article, "Reading in College," which he wrote as a senior. In it he deplored the reading habits of the students. "If a book is exciting they like it, no matter how false it may be to nature, or how openly it may violate the principles of art." As a remedy he proposed the following list for collateral reading, apparently in the order in which the names came into his head: Scott, Bulwer-Lytton, Cooper, Dickens, G. P. R. James, Thackeray's *Vanity Fair, Pendennis,* and *Book of Snobs,* Dumas, Eugene Sue, George Sand, Paul de Kock, Carlyle's *Sartor Resartus,* Emerson, Humboldt, Ruskin, Macaulay, Prescott, Niebuhr, Grote, Theodore Parker's sermons, De Quincey, Irving, Gibbon, Shakespeare, *The Spectator, Paradise Lost, The Divine Comedy,* Homer, Euripides, Aeschylus, Demosthenes, and Cicero. In asserting that this was not "a list of the books commonly read here," he probably intended to compliment his intimate friends.[3]

Incomplete as they are the lists indicate that the interests of Adams and his friends were distinctly literary and intellectual. The array of authors represents, however, a conventional taste, rather too preoccupied for the most part with accepted reputations. When they gave books as gifts, they gave them to

last for the ages, a very practical, but often uninspiring criterion. For example, Adams's generous Southern friend, James
May, gave him such works as: *Romans de Voltaire, Maximes du
duc de la Rochefoucauld, Pensées divers de Montesquieu, Oeu-
vres Choisis de Vauvenargues, Les Caractères de la Bruyère,*
and Inchbald's *British Theatre.* One hunts in vain for exotic
titles, for evidence of interest in literary byways, for some sign
of intellectual daring and originality. The enormous gaps are
eloquent of the coming triumph of the Genteel Tradition. *Hia-
watha* went from room to room, but neither *Moby Dick* nor
Leaves of Grass seems to have made any impression. There is
no hint of writers like Balzac or Stendhal or Villon or Rabelais.
Henry Adams eventually filled these gaps more successfully
perhaps than any other American of his time. That was to be
an important part of his life work.

Cambridge Life and Letters

Outside the classroom and the lecture-hall Harvard bustled
with life and offered an agreeable contrast to the "very retired
and quiet" life of Quincy, which had burdened Adams with
"care and ambition and the fretting of monotony." In 1855 he
became a member of the Institute of 1770, described by his
friend Nick as "a chartered literary society of great renown."
His maiden contribution of "three pieces, one in prose, one
poetical and one editorial" induced the secretary to record for
posterity: "These productions were excellent of their kind and
were received by the Society in a manner which must have
been highly gratifying" to the author.[1]

The Institute, which met regularly each Friday evening to
listen to the speeches and compositions of its members, fully
lived up to its literary reputation. A lecture on "Authors of the
Present Day" might be succeeded by an impassioned debate
on the question "Can We Justify Brutus in the Assassination

of Caesar?" or a harangue on the subject "Is Capital Punishment Beneficial?" Adams's lectures impressed the Secretary as "manly and forcible." Aware of his reputation as a contributor to the newly founded *Harvard Magazine,* the Institute elected him its editor, charged with the duty of overseeing the compositions submitted by the members.

On December 19, 1856, a few months after the more popular Nicholas Anderson was initiated into the Hasty Pudding Club, realizing the star of his "social hopes," Henry Adams became a member. In April of 1857 he took the part of Captain Phobbs in the Pudding's production of John Madison Morton's farce *Lend Me Five Shillings.* The playlet succeeded so well that Henry and his fellow actors repeated the performance at one of Dr. Johnston's private theatricals in Cambridge. Much luster was added to the occasion by assigning the women's parts to Cambridge girls. Crowninshield tersely commented that the performance was "fine." In the following winter Adams was "admirable" as "the argument-loving Sir Robert" in George Colman's *Poor Gentleman;* but the role he longest remembered and which most delighted him, as one of the shortest members of the cast, was that of the overbearing Captain Absolute in Sheridan's *The Rivals.* He appeared in at least one other play, *In for a Day,* his lack of brawn (he weighed only 125 pounds) qualifying him for the part of Mrs. Comfit. As many as forty honorary members of the society would come for the fun, and the young men were "incited to strenuous exertions," especially when the new professor, James Russell Lowell, was seen "applauding enthusiastically."

Adams's bookish inclinations were early recognized. By the end of his second year he was elected librarian of the Hasty Pudding Club. There devolved upon him and his friend Anderson the "Herculean job" of revising the catalogue of 6000-odd books. Adams's part of the task was "to re-arrange the books, re-fit and repair those which were dilapidated, and renovate

the whole system." Anderson, impressed by his friend's knowledge of books, praised Adams as a "modern Magliabechi,"
learnedly recalling the seventeenth century bibliophile. A further honor paid tribute to the librarian's satiric wit. He was
elected to the office of "Alligator" or "Krokodeilos" of Hasty
Pudding and he paid the customary forfeit of a piece of smok-
ing-room doggerel, still treasured among the arcana of the club.
Next followed his election as club orator. His chief performance
in this office took place at the semi-annual celebration of the
club, January 15, 1858. For some reason he felt the occasion
called for a homily. His oration, "The Fool's Cap and Bells,"
exhorted his hearers to avoid such follies as pedantry, literary
pretentiousness, and cynicism towards men's secret motives.
But worst of all follies, perhaps—and here he may have glanced
toward Concord—was the folly of impractical idealism which
sought "to regenerate the world and call it back from the hard,
selfish juggernaut track upon which it has trodden for these
three thousand years." The advice was sound enough by Harvard standards, but depressingly sober. Seventeen of the company, adjourned to Fontarive's French House for supper and
achieved "a general and exceedingly jolly and fiendishly noisy
drunk." This time they were lectured on their folly by three
watchmen.

From the far side of the Charles River, other energies than
alcoholic commonly exerted their attraction upon the mind of
young Henry Adams, for Boston was still the literary and cul-
tural capital of America. Plays, operas, concerts, orations, and
even raree shows competed for the patronage of young Harvard. In the journals and letters of Adams's friends we momentarily cross his path a hundred times going to and from Boston.
Perhaps he is in the company of young Crowninshield, puffing
a cigar on his way to hear Bendalari at the Melodeon. It may
be there are tickets to be bought for a concert of the Swiss
pianist Sigismond Thalberg or the German basso Karl Formes.

Now he is at the Boston Theatre where the drama glows "in its full splendor and purity." We hear his cry of admiration at Booth's performance of *Richard the Third*, "That man is acting as no man ever did before; it is magnificent, magnificent!" [2] There were other great occasions: Charlotte Cushman as Lady Macbeth or as Walter Scott's Meg Merrilies; Fanny Kemble reading *Twelfth Night*. More often of course when time hung heavy one had to be content with plays like *Who Stole the Pocketbook?* and *Retribution* or the minstrel antics of Perham's Ethiopians and the cavortings of the "Wild Man of Ceylon."

Visiting lecturers and orators provided very great models of the grand style. In December of 1855, Thackeray, at the peak of fame, lectured on "The Four Georges." Horace Mann, another of Adams's heroes, gave his satire on "Man." Far more impressive to college orators was Edward Everett's eulogy of Washington in June, 1857. They "thrilled with delight at the superb eloquence of the modern Cicero," whose "every sentence was a jewel in itself." When Thomas Hart Benton, greatest of the Western statesmen, lectured on the Union, even the laconic Crowninshield extended his comment: "There was a great deal of truth in his remarks." Adams heard an even greater master of oratory a few months before his own oration at the Harvard Class Day exercises. Rufus Choate delivered an oration on Hamilton and Burr. Among the dignitaries on the platform behind Choate sat Charles Francis Adams and "other wealthy men of Boston," seemingly "highly pleased" by the attack on trouble-making "republican politicians."

Among his contemporaries, Adams moved in the most select society. Anderson thought his dining club table "one of the nicest little assemblies in the city" and was delighted to join it. The chance of sharing a room with Adams filled him with anticipation of a cultivated existence. "What a pleasant, social time Adams, Dorr, and I will have next term," he told his mother in March 1856. "The rooms are as nice and cozy as is possible, and of a winter evening, how comfortable it will be to cock our legs upon the mantel-piece, and study or talk or read . . . We intend to pursue a course of reading, so as to keep up our knowledge of the Latin language as well as the French. We intend to have no idle hours; when not studying or reading some worthy work, we are to talk on various topics, and 'in primis' to argue and discuss all the subjects of the day, which, by the by, Adams is well qualified to do . . ." The unfortunate death of Hazen Dorr marred this idyllic plan; hence when Charles Francis Adams, Jr. was graduated in 1856, Henry took a room alone in Holworthy Hall.

The allusion to Adams's knowledge of "topics of the day" implies that Harvard did not weaken his interest in public affairs, once again *The Education* to the contrary notwithstanding. Study assignments may often have crowded politics to the back of his mind, but he was not isolated from the great world; if anything, Harvard was closer to the maelstrom than Quincy, where his father, who quietly labored in political retirement on the *Life and Works of John Adams,* had fallen "back into the narrow circle of New England life." [3] In his freshman year the aftermath of the Anthony Burns "rescue" violently excited the student body. Horrified by the honors which antislavery Bostonians accorded the unfortunate Negro, Anderson excitedly wrote his mother in Cincinnati: "Is this a civilized and enlightened community?" The law students at any rate agreed with him that it was not and they voted to censure the Board of Overseers because it had removed Judge Edward G. Loring as a lecturer on law for having upheld the Fugitive Slave Law. Adams like his fellow "Free Soilers" backed the antislavery board. In the realm of foreign affairs, the death of Czar Nicholas I in 1855 provoked interest in the progress of the war in the Crimea and students divided on the merits of the French and English pretensions.

Soon domestic concerns displaced foreign ones as news of the bloody Know Nothing riots reached the Yard. Did Henry share his friend's violent prayer: "Down with all foreigners who . . . oppose the rights of Native Americans"? Henry's grandfather, John Quincy Adams, had also dreaded the rising influence of "the Roman-Catholic Irish multitudes"; yet on the other hand it was the native American party which had overwhelmed the Free Soilers in 1854 leaving Charles Francis Adams "a leader absolutely without a following." In the face of such a dilemma Henry no doubt claimed the right of an Adams to form his own party.[4]

The political volcano began rumbling again during the election year of 1856 and erupted late in May. Senator Sumner of Massachusetts rose to speak on "The Crime of Kansas" and outraged his Southern opponents with harsh truth and equally harsh insult. If the truth was unanswerable, the insulting invective against Senator Butler of South Carolina could at least be avenged. Three days later, Butler's kinsman, Preston Brooks, strode into the Senate chamber where Sumner was at his desk and "beat him senseless with a stout cane."[5] This attack upon the man whom Henry Adams admired next only to his father gave him an intensely personal interest in the whole controversy, an interest complicated by the fact that Sumner might now be tempted to force his party into intemperate abolitionism. "The long agony of another Presidential contest," as Anderson described it, soon came to a climax and students hurried to Boston "to hear the returns as they came in" announcing the election of the evasive Democrat, James Buchanan. No sooner was Buchanan inaugurated than the Dred Scott decision furnished new fuel for controversy. But even such matters grew remote as the Panic of 1857 cast its shadow over Harvard and the young men faced the rigors of hard times parties. To Adams the suspension of specie payment in October had much more than a social interest, for his father had to meet the crisis as president of the Mount Wollaston Bank.

In short the period from 1854 to 1858 made as imperative a claim upon his political consciousness as any other period in his life and we may be sure that the tumult of these years filled the pages of the now-lost college diary. As he lectured classmates in 1858, "For four years we have associated together on the common ground of toleration; years during which the bitterest party spirit has excited the whole country, and stretched its influence even over us at times." And, only a few months after graduation, when he took stock of himself he clearly saw two traits at work in himself: "One is a continual tendency towards politics; the other is family pride." [6]

The Harvard Magazine

As an undergraduate, Henry Adams's response to the intellectual challenge of Harvard took characteristic form. Before the first year was out he was already contributing to the newly founded *Harvard Magazine*. The necessity for self-expression lived in the blood. From time immemorial, Adamses had worn out their lives at their writing desks, conscientiously transferring their views of a refractory world to the endless pages of their journals. One of Henry's first acts upon entering Harvard was to begin keeping his own journal, as his father had done thirty-three years before when he entered Harvard. His brother Charles's journal was already in its third year. There is something almost of pathos in the picture of the two brothers in their room at Mrs. Storey's, entering each day's epitaph with the sense of a hypercritical posterity—the fate of an Adams—standing at their shoulders. Of Henry's journal nothing survived the impulsive destruction of his papers that followed the suicide of his wife in 1885; of Charles's diary only a little was salvaged in his *Autobiography*.[1]

Even from the beginning Henry's writing had guidance of a sort, for Charles had an affectionate interest in his progress.

The intellectual intimacy between the two brothers was already well established as we can surmise from the mood of their early letters. At the opening of their correspondence in 1858 we break in upon a long standing debate in which Charles insistently urges greater literary activity and Henry humbly acknowledges: "I stand in continual need of some one to kick me, and you use cowhides for that purpose." [2]

Formal instruction in rhetoric did not begin, however, until the second year and then the scholarly professor Francis J. Child showed Adams the importance of writing "clearly, forcibly, and harmoniously" and of using "striking observations, original metaphors, or pertinent similes." In his assignments of themes and forensics Child showed no interest in his students' opinions on contemporary affairs, but usually required them to base their papers on solid reading. For one theme Adams dug in Voltaire's *Charles XII;* for another the trail led to Milman's *History of the Jews.* One paper analyzed the character of Horatio; another answered the question: "Was Shakespeare Indifferent to Fame?" Perhaps like Anderson, Adams thought that question an "absurdity of absurdities." A paper on Goldsmith's "Deserted Village" required careful exegesis of the last few lines of the poem. A more abstract line of inquiry was suggested by such topics as "The Errors of Great Men" and "What Design of Providence is Accomplished by the Universal Propensity to Build Castles in the Air?" One of the most challenging forensics dealt with Oliver Cromwell, whom Adams regarded as an early saint of democracy. None of these juvenilia appear to have survived, but, as we have seen, Professor Child was very favorably impressed.

The *Harvard Magazine* was founded in the autumn of 1854 by a group of earnest young men headed by Phillips Brooks and Franklin B. Sanborn, the future biographer of Thoreau. Sixteen years had gone by in which no literary magazine had been published at Harvard. Imitating the genteel reticence of the literary quarterlies, the editors published all contributions anonymously; but this must have been more fashion than modesty, for the authorship of the articles was always an open secret.[3] In their front-page prospectus the editors nailed the following defiant invitation to the masthead: "In the discussion of all subjects we shall aim to give the greatest freedom and invite the most opposite opinions." On the back page, however, there appeared the discreet second thought: "Our magazine is started with no intention of using its pages to 'squib' the College government, and we shall avoid personalities of every kind."

By the middle of the second semester Adams's first article was in print, a fanciful historical sketch called "Holden Chapel," in which that storied piece of Harvard architecture served as a peg to hang four reverent vignettes of Revolutionary times. Its sole interest lies in its foreshowing of two preoccupations of his subsequent literary career: the depiction of character and concern with the fortunes of his ancestors. "Here we have a man now, who looks as though he were Governor of the Province, so richly is he dressed, and so much respect is paid him by the others. And there is another, dressed very plainly, but with a firm look upon his face as though he would persevere in his object in spite of the King of England himself . . . And now a third man approaches. He is short and stout, apparently yet young, and quite good-looking. Mr. Hancock returns his salutation and says, 'Well, Mr. Adams, how comes on my lawsuit?'"

"My Old Room," his sophomore's farewell to Hollis 5, is a refreshing piece of autobiography, having no trace of his later abhorrence of the pronoun "I." His room was "the coldest, dirtiest, and gloomiest in Cambridge," but, he went on, "To me it will always be haunted by my companions who have been there, by the books that I have read there, and by a laughing group of bright, fresh faces, that have rendered it sunny

in my eyes forever." Recollections like these taught him that
it was not a Harvard education that was at fault but rather the
"respectable" student whom he thus sketched: "A little reading,
a little study, a little smoking, a glass of wine occasionally, a
select acquaintance which frowns gently on vice, and various
homeopathic doses of negative virtues." He protested that "dissipation is the exception, and not the rule," and likewise rejected the common imputation of atheism. "If at any time stray
professors of infidelity have come among us, their opinions
have arisen in another soil, and have found nothing kindred
to themselves." At the same time, he confessed that although
the two "years have been spent very soberly, and indeed very
profitably . . . neither I nor many of those who came here
at the same time with me have used them wholly as we ought."
Furthermore, he discovered something that was to make him
uneasy for the rest of his life: "I have learned too, however
late, that College rank is *not* a humbug, as some pretend."
Against the vulgar opinion that "college is dangerous and hurt-
ful," he retorted that "the store or the counting-room is as bad,
or worse," thus arraying himself on the side of the angels and
of his family in its immemorial quarrel with State Street.

The desire to improve his fellow students continued to agi-
tate him, and in the following year he returned to the attack
in his essay "Reading in College" and again reminded his fellows of their intellectual and moral debt to Harvard. "The
most vulgar mind and the poorest understanding must be somewhat elevated and instructed by two years of contact with what
is better than itself." Adams disclaimed any wish to be a censor of manners, but the disclaimer lasted only until the publication of his next article, "The Cap and Bells," which consisted
of moralizing extracts salvaged from his Hasty Pudding Club
oration.

His most coherent article, "College Politics" (May 1857),
touched upon the chief controversy of his college career: the

corrupt influence of the Greek letter societies.[4] Its roots went back to the day when his grandfather persuaded the Phi Beta Kappa chapter to abandon its secret rites. Henry's father bequeathed the family horror of secret societies in words like those prefixed to his edition of John Quincy Adams's *Letters and Addresses on Freemasonry:* "An obvious danger attending all associations of men connected by secret obligations, springs from their susceptibility to abuse in being converted into engines for the overthrow or the control of established governments." Like his father Henry believed that "secret combinations of men, which attempt to exercise a political influence upon others by means of their organization, are bad in theory and practice." In the portentous style which he clung to for many years he warned that "Within the last few months, events have taken place which threaten to make a deep and lasting change in our social relation." The rhetoric was embellished with an inevitable allusion to the Hydra whose many necks required severing.

The most feeble of his *Harvard Magazine* contributions, "Retrospect," filled with sententious and sentimental meanderings on the subject of vacation and angels "in muslin," has an unintended value to the modern reader because of the hint it gives of the frustrations that sometimes clouded his boyhood in Quincy where two elder brothers ranked before him. "The time is not slow in coming when we want to be back with our friends again; to be back in the delicious independence of student life; to be back where we are not lost in the multitudes of the world. Even home becomes irksome. Ennui is fearful. It is mortifying to associate with our elders with whom we have no common tastes or ties." Evidently, there were times even when his father's self-righteousness exasperated him. As he confided to Charles, "God Almighty could not get an idea out of his head that had once got in."

A remaining article, a spirited *jeu d'esprit* jibing at Thomas

Keightley's new edition of the *Fairy Mythology,* is similarly valuable. "If to be consoled by the faults of others makes a man a dunce, we acknowledge the infirmity, and are not afraid of the epithet." Expressed thus for the first time is the self-protective formula which was ever after to compensate for his sense of insecurity and inferiority: One's own failures are excusable if better men fail as badly. Six months later he excused himself and his classmates in these words: "But though we acknowledge our own ignorance, we know that every one else is ignorant also." By the time of *The Education* the formula had become a fetish.

The Pattern of Success

Having made his mark in the *Harvard Magazine,* Adams set about in his last year at college to vindicate his reputation as a scholar. He decided to compete for the Bowdoin Prize, an annual literary contest in which his brother Charles had won first prize a few years earlier. Henry achieved second place and the consolation of $30 prize money. Notable as the triumph was, it may well have seemed to him a melancholy proof of the due order of promotion in his family.

The contest offered a choice between two subjects: "An Examination of the Evidences Historical and Traditional, for the Existence of Robin Hood" and "Saint Paul and Seneca." Being currently preoccupied with the ethical systems of Bowen and Jouffroy, Henry chose "Saint Paul and Seneca." This first extensive historical composition exploited a kind of history to which, fortunately, he never again returned. It was history in the vein of Plutarch, history as his grandfather preferred it, a "school of morals." [1] In working out the parallel between the Christian apostle and the pagan moralist, Adams relied heavily upon Conybeare and Howson's *Life and Epistles of St. Paul* and G. H. Lewes's *Biographical History of Philosophy.* The

form followed that of Plutarch: first came the life and character of Seneca; then the life and character of St. Paul; then a
parallel life contrasting the two characters very much to the
advantage of the Christian St. Paul.

What was especially notable was the highly sculptured character of the prose, filled with rolling Ciceronian periods, an extreme example of what he soon called the "stilts" of his college
writing. The stately antitheses were here and there ornamented
with classical similes: "Like the signal fire that announced to
Clytemnestra the capture of Troy, it flashed from Mt. Ida and
Mt. Athos, and bounded over the plain of Asopus even to the
Saronic gulf and to the roof of the Atridae." In his moralizing
conclusion he asked cynics to contemplate history "and say if
they can that the world is not infinitely better and happier than
it was."

Of all his college compositions the one that most fully reflected his romantic idealism and his abiding spiritual hunger
was the address which he read as Class Day orator on June 25,
1858. The memorable day arrived with a temperature of ninety
in the shade and not a breath of air stirring among the elm
trees of the Yard. Shortly before noon, after a collation of ice
cream, the Class of 1858, the faculty, and the crowd of guests
marched from the President's House to the First Parish Church
to hear the handsome and earnest looking spokesman of the
class who mounted the platform garbed in the traditional black
gown. He was surprisingly short in stature, barely five feet
three inches, but the impression faded before the searching
glance of his dark eyes with their reminiscence of president
ancestors.

The grave young orator began with a disarmingly academic
review of the history of Class Day celebrations and then by
degrees he entered upon an apologia for the senior class, as if
to protect its members from the trite exhortations of the bac-

calaureate sermon that was soon to come. "We are not original nor geniuses, but only an ordinary set of young men with fair acquirements and average abilities, who have got to hammer out their fortunes as well as they can."

Finally he began to sound the main theme of his oration: the dangers of materialism and the commercial spirit. He scored the low ambitions which so many of his classmates professed. Unwittingly the study of moral philosophy had fostered that spirit because it taught "how false the hopes of the young are." (We can imagine the frowns upon the faces of President Walker and Professor Bowen at this sally.) Had not "all philosophy from Solomon to the last number of *The Virginians*, been singing that old song, *Vanitas Vanitatum*"? And yet, their course in political economy which professed to be a sister science of ethics seemed to enjoin upon them that chief vanity, the duty of amassing wealth. How could one accept the paradox of a society in which men were unintentionally moral? He further questioned whether in following "in the footsteps of Thackeray and Carlyle" in warring against "shams and formulas" his classmates had not come to "laugh at things that were not shams"; for students "are fond of burning other people's idols." (His friend Anderson must have squirmed at the thought of his once expressed contempt for "those grand old humbugs, the Puritans.")

There were a few, however, who still held to their scruples. "Some of us still persist in believing that there are prizes to be sought for in life which will not disgust us in the event of success . . . There are some who believe that this long education of ours, the best that the land can give, was not meant to be thrown away and forgotten; that this nation of ours furnishes the grandest theatre in the world for the exercise of that refinement of mind and those high principles which it is a disgrace to us if we have not acquired." These students, he concluded, "will still put their whole faith in those great truths, to the advancement of which Omnipotence itself has not refused its aid."

The study of past ages and the classics had helped to show him the limitations of modern science. "It is puffed up by its self-conceit because a certain Francis Bacon, some two centuries ago, happened to put it upon the right track and it has run along downhill ever since." (Professor Agassiz must have solemnly nodded agreement.) He and his classmates had grown aware of the vast extent of man's ignorance. The study of mathematics, chemistry, and metaphysics had taught him "that though man has reduced the universe to a machine, there is something wanting still; that there are secrets of nature which have puzzled chemist and philosopher even in these days of science, and which still wait for a solution; . . . that there are problems in the relations of man to man which political economy has tried to solve again and again and never succeeded; that there are questions in the relation of man to the universe and to the future, which all metaphysicians from Thales downwards have worn away their whole lives in striving to answer, but have failed, always failed."

It was the sermonizing of a deeply troubled young man, Puritan to the core in his contempt for the world as he found it. Harvard had helped him discover the immense chasm separating the moral order from the base contradictions of the modern world. Neither man's reason nor his science held out much promise of bridging that chasm. The hope of the world must therefore lie in faith in the supremacy of the moral order.

The oration gives striking evidence of the deeply spiritual, even religious bent of young Henry Adams. Especially significant is the fact that he chose to talk on what was an unusual theme for a Class Day celebration. Like his oration before the Hasty Pudding Club this one seems to have been equally inappropriate to the occasion, for custom required an oration on the pleasures of college life. The compulsion that drove him to place the main emphasis on the spiritual crisis of mankind, it can now be plainly seen, was to dominate his thinking to the end of his life.

In an editorial in the Springfield *Republican* eleven years later, his father's friend Sam Bowles recalled that the oration was "distinguished for little except its irony and cynicism." Pessimistic, it may have been, but hardly cynical. Two contemporary comments acknowledge its merits. His friend Crowninshield, who had been busy all day with the Class Day arrangements, hastily jotted down, "Very good." His father, Charles Francis Adams, entered in his journal for the day: "Henry has been noted for his faculty as a writer which is the thing that secured him this honor, and he sustained himself both as a speaker and a writer on this occasion."

Among the crowding literary activities of the last term, there was one final bit of writing that can also be set against the disenchanted mood of *The Education*. This was an autobiographical sketch which tradition required Adams to contribute to the huge "Life-Book" of the Class of 1858. In the scant six hundred words he put aside all literary pretensions and wrote with becoming candor: "I have had an infinitely pleasanter time than I ever had before . . . I do not believe it would be possible to pass four pleasanter years." He was unaffectedly grateful for his election as Class Orator. It was "a compliment which pleased and gratified me more than any other which I ever received, or, probably, ever shall receive."

As for his plans for the future, he wrote: "Having now come to the end of the Course, and feeling no immediate necessity of making money, it is probable that I shall soon go to Europe, where I hope and expect to work harder than I have ever worked at Harvard. My immediate object is to become a scholar, and master of more languages than I pretend to know now. Ultimately it is most probable that I shall study and practise law, but where and to what extent is as yet undecided. My wishes are for a quiet and a literary life, as I believe that to be the happiest and in this country not the least useful." The fulfilment of those wishes was to be long deferred, but ultimately they were to be richly granted.

Chapter Two

The Grand Tour

1858–1860

The Pierian Springs of the Civil Law

THE LONG SUMMER DAYS at Quincy that followed the College Commencement of July 21 slipped away, leaving no memorial beyond Adams's catalogue of his personal library. Important family changes, however, already impended; his father's long retirement from public life was coming to an end. Thanks to the influence of Senator Sumner, Charles Francis Adams was nominated on October 7 as Republican candidate for Congress from the Third Massachusetts (Quincy) District. Despite some shaking of heads over the unpopularity of the former Free Soil Party candidate he swept the district by a vote of more than two to one.[1]

Young Adams had gone ahead with his plans, finally fixing on a rigorous course of study to occupy two years. First he would master German and then attend a course of lectures on the Civil Law at the University of Berlin, at the same time engaging a Latin tutor and learning to translate Latin into German. Afterwards he expected to study in Heidelberg and possibly Paris. Others of his classmates were similarly busy with their plans; for, as Anderson reported, "Quite a deputation will be sent from my class to polish off in the elegant cities of Europe, and obtain that knowledge of the modern languages so necessary to the present age."[2]

Adams's decision to study Civil Law at the University of Berlin was demonstrably a sound one, whatever it may have looked to him in retrospect. There, in 1810, Savigny had founded the modern study of Roman law and there his pupil Georg Puchta, whose *Cursus der Institutionen* became Adams's vade mecum in his travels about Europe, introduced the scientific study of the subject in the seminars for which the university was growing famous. As for the value of the subject itself, Henry and his father had every reason to know what they were about, though long afterwards Henry denied it. When the edition of John Adams's *Works* took form between 1848 and 1856, father and son read in the diary of their first president how at the age of twenty-two he began the study of Justinian's *Institutes:* "Few of my contemporary beginners in the study of the law have the resolution to aim at much knowledge in the Civil Law; let me, therefore, distinguish myself from them by the study of the civil law in its native languages, those of Greece and Rome." At forty he advised his young friend Jonathan Mason: "You should go to the fountainhead, and drink deep of the Pierian spring. Justinian's *Institutes,* and all the commentators upon them that you can find, you ought to read. The Civil Law will come as fast into fashion in America as the French language, and from the same causes." He had obviously been over-sanguine; nevertheless professional interest in the subject did grow and at the dedication of the Harvard Law School, President Josiah Quincy urged the need for thorough training in the Civil Law.[3]

Moreover, Henry and his father may well have considered the role that a knowledge of the Continental codes was likely to play in American legislation. The movement toward the codification of state laws was already under way, New York having initiated the trend in 1846. Lawyers were also beginning to see the importance of uniform federal laws and the need for a more scientific approach to legislation. To a prospective lawyer-statesman, to whom the everyday practice of commercial law must be a necessary evil, such larger prospects would be decidedly attractive. But even these considerations must have been secondary to the immediate influence of Senator Charles Sumner's example. To prepare himself for his career at the bar, Sumner had spent three years in Europe in hard study of Continental government and jurisprudence, and his legal erudition was unmatched by his contemporaries. As a close intimate of the family, he inevitably discussed the value of European study with young Henry who had been his admiring disciple. All things considered then, an ambitious young man could have done much worse than propose to himself the study of Civil Law.

Part of the "deputation" of gilded youths, Adams among them, left Boston on September 28 and sailed from New York harbor on the following day aboard the luxurious sidewheeler, the "Persia." The party included such congenial classmates as Nicholas Anderson, Benjamin Crowninshield, and Louis Cabot, but there was probably little appetite for merriment as the eleven day passage to Liverpool proved rough and tedious. Prone to seasickness as he was, Adams was no doubt wretched most of the time. After a few days in London he crossed to Antwerp with Crowninshield, stopped off at Hanover where he left his companion, and proceeded alone to Berlin.[4]

Upon reaching Berlin on October 22, Henry was delighted to learn that the "Governor," his father, had won the nomination on the very first ballot and the subsequent news of the election made him jubilant. "The old Free Soilers, sir, are just about the winning hosses, I reckon, just now," he wrote to Charles.

His elder brother had recently completed his apprenticeship as a law clerk in the office of Richard Henry Dana, one of the family's oldest intimates, and had been admitted to the bar. While waiting for paying clients, he now turned much of his energies as a counsellor toward his younger brother Henry whose carefully laid plans had gone awry from the beginning.

Henry justified his original plan of study as being "simple enough; useful enough; and comprehensive enough." But he soon realized he would not be ready to join the university until the middle of the term, certainly not before January. Besides even if it were wise to break "in on a course of lectures" he could scarcely hope to become a student of the Civil Law unless he were "an absolute master of written and ordinary Latin." For a while, therefore, he sedulously grasped at straws and gravely floundered among alternative projects.

The German language proved intolerably difficult. "I can't once in a dozen times speak a grammatical sentence, and understand what is said only when very slowly spoken," he complained. "As for a continued lecture, I can't catch anything at all, and at the theatre very little." At most he was able to endure only a few visits to the University of Berlin, perhaps dropping in on Professor Heinrich von Gneist's course of lectures on the Roman law, or Professor Leopold von Ranke's more famous lectures on history. How great the difficulties were we may gather from a comment of young Crowninshield, who was a considerably better linguist than Adams. Von Ranke "gulps his words" so that "not a single idea can be followed through." In the face of these obstacles Adams turned with fairly good humor from a direct assault upon the Civil Law to a more prudent campaign against the languages. Putting pride aside he enrolled in the Friedrichs-Wilhelm-Werdesches Gymnasium. The university experience grew in the telling, however, and the Class of 1858 secretary, taking the promise for performance, reported that Adams had been "attending lectures in the University." By 1888 the class historian had amplified the legend to "two years in Germany, much of the time at the universities." [5]

Henry tried to satisfy his brother's doubts about the wisdom of his current studies by outlining his long term plans. After Europe he proposed to return to Boston to study law there for two years and then emigrate to St. Louis. "What I can do there,

God knows," he exclaimed. "But I have a theory that an educated and reasonably able man can make his mark if he chooses . . . I must, if only I behave like a gentleman and a man of sense, take a position to a certain degree creditable and influential, and as yet my ambition cannot see clearly enough to look further." He admitted the claims of family pride; but he was repelled by the humiliations of a political career. He was determined, he said, not to "quit law for politics without irresistible reasons."

With sharp insight Charles thought his younger brother unfitted for the law; and he pressed him to turn at once to writing for publication so as to bring his name before the American public. The alternatives with which he tempted his client ranged from newspaper writing to "greater literary works," such as novels and histories. Henry gloomily acknowledged that he felt "as certain that I shall never be a lawyer, as you are that I'm not fit for it"; at the same time he felt himself unprepared to attempt a pretentious literary work. Yet that was "the only one of the branches" of Charles's suggestion that struck him "as practicable." Momentarily he grasped at the idea "of editing our grandfather's [John Quincy Adams] works and writing his life," a task which might be "enough to shape his whole course." On second thought, however, he shrank from even this prospect, declaring that "it is not in me to do them justice."

Thus at twenty he despaired of his own powers. "I am actually becoming afraid to look at the future, and feel only utterly weak about it . . . This is no new feeling; it only increases as the dangers come nearer." It is evident from the echoes in Henry's letters that his brother expected exceedingly great things of him, expected him to combine "the qualities of Seward, Greeley, and Everett" as he protested. He vehemently disclaimed such powers: "Mein lieber Gott, what do you take me for? Donnerwetter! . . . I know what I can do, and I know what a devilish short way my tether goes." Even though the law was "not a pleasant study," and though they were "not adapted to make great lawyers" the alternatives were worse. "You would make me a sort of George Curtis or Ik Marvel, better or worse, a writer of popular sketches in magazines; a lecturer before Lyceums and College societies; a dabbler in metaphysics, poetry, and art, than which I would rather die, for if it has come to that, alas, verily, as you say, mediocrity has fallen on the name of Adams."

Law, at least would provide an income and a springboard to a public career, as in the case of Everett, Sumner, Palfrey, and their own father; whereas, there were few "literary pursuits that produce money," that is, pursuits for which he was eligible. Newspaper editing was as open to objection as law. As for writing for the *Atlantic Monthly, Putnam's,* and *Harper's,* "rather than do nothing but that," he witheringly remarked, "I'd die here in Europe." He insisted, however, that their ideas of success were not really as far apart as Charles seemed to think; it was merely that Henry desired to avoid unnecessary hazards. They were both in approximately the same case, and "beautifully adapted to work together." Neither one of them, he cautioned, ought to take any step which might "knock one of us in the head forever, or so separate us that our objects would become different." Charles's error lay in assuming that Henry was "capable of teaching the people and becoming a light to the nations"; he, on the other hand, was willing to leave that possibility to the future and not risk "a public disgrace from slumping" as he would "infallibly" if he now acted on Charles's assumption.

Ignoring these morbid anxieties, Charles offered to find a market for any article that Henry should care to write. Under pressure Henry admitted that "it would not be impossible to write an article on the Prussian schools" now that he was enrolled in a German Gymnasium. "The subject, as I would treat it, offers a pretty wide surface for anything that I should care to say, whether political, metaphysical, educational, practical, or any other 'cal.' " Having reluctantly ventured this half-promise he returned to biding his time.

As the winter passed, he faithfully grubbed away at the study of languages among his young classmates. These had such "perversely wrong" notions of American life that he amused himself, as he told his distinguished correspondent, Charles Sumner, "by giving them original and astounding ideas of my own. Indeed I expect that in a short time they will really believe that I am an Indian with two squaws and corresponding papooses and live in a wigwam adorned with scalps." To Sumner, who showed a friendly interest in his studies, he sent a respectful report of progress: "I am in school every day from three to six hours, and generally have to come at eight o'clock in the morning. Yesterday I translated into German a page of Xenophon, before the class, and today in Caesar." For his own pleasure, he added, he was reading Ovid's *Metamorphoses* and Kugler's *History of Painting.* To his brother who approved the Gymnasium course somewhat too warmly for Henry's comfort, he adopted a strongly deprecatory tone: "You estimate the effect of school too highly however. It has enabled me to give method and concentration to my studies, but I have found here that it is impossible to go back ten years in one's life . . . My *r* is already formed in a very different way, and the process has very little influence on me." [6]

By early March 1859, the article on the Prussian schools with which he had toyed back in November, had passed through the research stage in which he explored many works on "das preussiche Schülwesen." Within a few weeks, it had grown to "enormous length," but his dissatisfaction grew with it. Not until May did he finally finish recasting the troublesome composition as "Two Letters on a Prussian Gymnasium." It still seemed to him "very poorly written and excessively stupid," and to his discriminating taste it was in a "wholly unpublishable state." The self-judgment was understandably violent for in this first try at professional writing his eyes were opened to the difficulties which he would have to surmount in making his way by the pen. Having broken down on the point of form the article failed to rise much above the diffuse and repetitious style of his college efforts.

What the unpublished twenty-two page manuscript may lack in literary excellence, however, is offset by its autobiographical interest. Horrified by the squalor of the German classroom, he recalled his own school days with an entirely new pleasure, especially the boys with whom he used to go to school, their animation, their well scrubbed faces, and their tidy clothes. When he looked at his pale, badly nourished German classmates, he was swept with homesickness for the sight of his healthy schoolfellows as they used to crowd into the classroom glowing and breathless after wholesome play. Even though he did not believe that his own school days had been the happiest days of his life, he nevertheless now thanked Heaven that he had not been educated under what people unthinkingly called "the most perfect school system in the world."

Warm hearted and friendly, Adams became the champion of the smaller youngsters against the bullies of the class. Often they showed their gratitude by cramming some of their breakfast of black bread into his mouth as they climbed all over him and rode upon his back. But sometimes when he was pestered by impertinent questions about America there were limits to his indulgence; then he would tease the children with fabulous stories or if greatly provoked, as once by a Jewish boy, he would box his questioner's ears.

If Adams was dilatory about the "Two Letters on a Prussian Gymnasium," the same thing could hardly be said of his letter writing. Perhaps mindful of his father's maxim that "there is no species of exercise, in early life, more productive of results useful to the mind, than that of writing letters," he practiced

that art with the enthusiasm of a young Horace Walpole. For
a time he wrote at least one long missive daily, though he did
not often reach the four-thousand-word length of one of his
letters to Charles. Sometimes he got himself into difficulty for
his habit of laying things on with a trowel, as when his brother
brought him up sharply for some cutting observations about
Boston girls. On that occasion Henry reminded him that, after
all, letters were reflections of the passing moment. "It is not
strange," he said, that "when I am hurrying to put down the
first thing that comes into my head . . . I say many silly
things." The "silly things" were hardly serious blemishes and
Henry was quite aware that his letters, however impromptu
they were, were no ordinary compositions. Of some he kept
copies that he sent on to Charles for safekeeping.[7]

Exposed to a new current of experience, meeting people who
had none of the inbred taint of Boston about them, and free
to do as he pleased, Henry filled his letters with the full savor
of his new life. "I always had an inclination for the Epicurean
philosophy," he told Charles, "and here in Europe I might
gratify it until I was gorged." Life was not "wildly exciting"
in Berlin, burdened as he was by his studies. Yet often enough
he yielded to the great temptations of the opera where the
"glorious" orchestra, scenery, and ballet carried him more or
less safely across the abyss of the language and to a lesser extent he patronized the theater. Mozart's *Zauberflüte* and Bee-
thoven's *Fidelio* taxed his understanding somewhat and *Hamlet*
in German sounded rather "flat," but Fräulein Taglioni's ballet
dancing was unalloyed and unforgettable pleasure. Sometimes
his social activities seemed to him too tame. "I've not been on
a real bat for ever so long," he would reflect. "I'm virtuous as
St. Anthony and resist temptation with the strength of a mar-
tyr." Anderson, who was also settled in Berlin, lived "a long
way off" and Adams saw him only occasionally. For companionship he counted chiefly on James J. Higginson, one of Charles's

classmates, in the main avoiding compatriots so that he might establish German connections. In this project he was disappointed as there was little student life and that was "dirty and fleay." Occasionally there were agreeable interludes and though a bottle of wine was the "outside" of what he could carry, he would make it a point "never to refuse a good glass when it's offered." There were therefore moments when he found himself flushed with wine, the inevitable cigar in his hand, and "talking very fast."

In mid-March his friend Ben Crowninshield came up from Hanover and for the next three weeks the young men saw each other daily. Crowninshield was struck by the beard which Adams had grown since he last saw him. He also noticed that his friend had grown "a little fatter," and this in spite of the execrable diet. They passed their time in walking about Berlin, visiting concert halls, museums, theaters, and wine cellars with limitless energy. Adams did not feel wholly comfortable, however, in the un-Puritanical atmosphere of Berlin. "Fellows who can live on music or art or women are all very well here," he explained to Charles; "I've done as well as I could at all three. The two first are good. The last is a damned humbug."

On April 12, 1859, at the close of the first semester of the Friedrichs-Wilhelm-Werdesches Gymnasium, Henry cheerfully cleared out of Berlin and departed for Dresden in the company of Crowninshield, Higginson, and the Boston Apthorps. His lively letter describing the journey by way of Wittenberg and that "funny little Pumpernickel" of a town—Dessau—bubbles with gaiety. Especially delightful is his story of the walking tour through the Thuringian forest, the madcap threesome having now become four with the arrival of young John Bancroft, the historian's son. In the snow swept mountains Ben Crowninshield's "real ten-horse power" Tom and Jerries "had a miraculous effect" on their spirits, but on the third day the four companions gladly compromised on a wagon and whiled away the time "in an intellectual and highly instructive series of free fights to keep us warm, which commonly ended in a general state of deshabille all round."

In Dresden Adams's plan to study the Civil Law soon became little more than a nominal occupation. He did no more than "pretend to read a page of law a day, an effort which unhappily never succeeded," and he now saw he would do little "real work" in Europe. He continued to improve his German by reading the learned Puchta, but deferred more serious study until he should return to Berlin in the following November. He was now frankly achieving "a gloriously pleasant and lazy condition." For the next several months he and Crowninshield were almost inseparable and his friend's diary is a record of constant sightseeing and theater-going. They attended performances of *Oberon, Rienzi, Don Juan, Tannhauser,* the *Prophet,* and such ephemera as *Berlin, How It Weeps and Laughs.* Dutifully they presented themselves at the Belvedere concerts, visited the museums and art galleries, and toured the medieval churches of the locale. When Crowninshield was indisposed for a few days Adams read himself hoarse from such current enthusiasms as Schiller's *Thirty Years' History,* Carlyle's *Frederick the Great,* or the novels of Paul de Kock, in German. Of course a staple matter of interest was German grammar, in which Adams was still not proficient in spite of the conversational opportunities of his *pension.* One day he finally agreed to Crowninshield's plan to carry on their conversations in German.

Few things marred the tourist delights of the two Americans. True, rumors of war had filled the papers all winter and bloody fighting quickly followed the mobilization of troops in Austria, France, and Piedmont in the early spring; but in Berlin it had all seemed so remote that Henry could not conceive its affecting his plans for travel. "It isn't probable that any Austrian will shoot me in the valleys of the Tyrol." However, war grew to be an important topic of conversation in Dresden that spring and it added a dubious zest to his life. Professing to be neutral, he found it "deuced hard to avoid cheering" when news of the great victories at Solferino and Magenta came in. The fact was that he strongly sympathized with the Italians and their French allies in the Italian struggle against Austrian tyranny. His conversational formula for escaping embarrassment was simple: *Frankreich und Oesterreich sind mir ganz einerlei* (France and Austria are all one to me).

On the last day of June the young men set out together for an extended tour of Bavaria, Switzerland, and the Rhine country. The sightseeing went on strenuously until mid-August when Henry joined his sister at Bern to go on to Italy with her, the signing of the Treaty of Villafranca on July 11 having made travel safe. On September 9, after his return, the two friends resumed their tour. Down the Rhine they went en route to Cologne. The fabled haunt of the Lorelei "spoiled the Rhine" for them for it turned out to be nothing but "a little tame rock." At Cologne Adams caught his breath long enough to fill the gaps in his journal and then the travelers pushed on to Louvain, Brussels, and finally Antwerp, omitting no important cathedral nor celebrated painting en route. They had now come full circle in their travels, for Antwerp was the first European city they had seen on their arrival a year before. In the distance the lovely tower of the cathedral reminded them agreeably of "the lighthouse to the Port of Boston." After lesser cathedrals, that of Antwerp seemed to them "the perfect example of the art of architecture." Carefully studying their Murray and Baedeker they went on to Rotterdam, mustered up thirteen gulden to hear the great organ play at Haarlem, and re-entered Germany by way of Amsterdam. After the freedom of Belgium and Holland, the "old oppression" of Germany was loathsome to them. As Berlin still seemed depressing, they promptly left for Hanover and a visit to Hildesheim. Here after a day of sightseeing with the usual inspection of the local cathedral the young men explored the wine cellar of the Dom Tavern where for the first time in their travels Adams completely forgot Boston. Maraschino followed "the true nectar" of red wine and "Old Spanish," followed the Maraschino, until as Crowninshield phrased it, "we were, as was natural, a little gay of spirit." Adams wandered off mistily into the night, walked into a strange house and insisted on making his bed on a convenient trunk, over the protests of a bevy of young women. After an amusing and harmless contretemps the friends were safely reunited.

They got back to Berlin, paid their respects to Minister Wright and busied themselves with plans for a studious winter. The plans, however, were no match for chilling mid-October weather which gave Adams a depressing foretaste of greater discomforts to come and inspired in him "new and stronger outbreaks against the cold, unfriendly, wet, soulless town." In his disconsolate mood he looked "like a beaten dog with the tail between the hind-legs." Unwilling to face the cold horror of Berlin in wintertime, he wrote to Vienna and Dresden hoping to be taken in by a well disposed family who might further his German studies. On October 24, 1859, welcome word came from Madam Reichenbach, wife of the geologist, that she would take him into her family in Dresden, as earlier she had entertained his admired teacher James Russell Lowell. Gratefully Adams made his way back to Dresden, all thought of study at the University of Berlin extinguished at last.

He settled down again to a regular program, spending three mornings a week with his fencing master and three with his riding master. German still continued to be his main object and he pursued it in books on the "constitutional history of various countries and desultory light reading," a program varied as usual by tourist pursuits, including tours of the geological museum under Herr Hofrath Reichenbach's expert direction. It was a delightful program but his pleasure was dampened by letters from home.

Not for long at a time was he allowed to forget that he was the first of the family to be indulged in the Grand Tour. Weighed down by the sense of being "under obligations and bonds for future conduct," he sometimes felt he had got hold of an indigestible "big plum cake." Ever in his mind was the prospect of the inevitable accounting to his father who, he was fully prepared to hear, would "lay the fault of every failure and every error in my life to Europe." The tenor of his father's letters had not been indulgently understanding. Even Charles refused to be put off easily. Quick to see that Henry's plan of study had come a cropper, he did not hesitate to call him a "humbug" and to reproach him for wasting his opportunities. With some petulance Henry accused Charles of "smouldering worse than I" and neglecting Boston "literary society whom it would be well worth while to know, beginning with Waldo Emerson and going down." In his own defense, he asserted that their father had advised him "against writing magazine articles on the ground that they are ephemeral."

Beneath this contest of purpose there lay of course the profoundly unsettling hope of the family, the hope of their father's "living in the White House some day." Henry feared, however, that the onset of the "irrepressible conflict" might blow them into the howling darkness of opposition politics. His father, now in Congress, having begun to incline toward Senator Seward and his middle-of-the-road policy on the slavery question, was veering away from the irreconcilable Sumner. Fortune would eventually smile upon them all only if Seward could be "quietly elected President of the United States." Not that Henry had abandoned the antislavery cause, but rather that he "would thank God heartily to know that comparatively conservative men were to conduct this movement and control it." Already

he had made his own boyish effort to protect his father and
Seward from the influence of the Radical monomania which
had begun to possess Sumner's mind.

More than three years had passed since the brutal attack
upon the Senate floor, yet Sumner had not yet recovered from
its effects in spite of European travel and heroic medications.
Southerners who had exulted in the savage conduct of Preston
Brooks now high-mindedly accused Sumner of shamming illness "for political effect." His failure to resign his Senate seat
lent color to the charge and thus embarrassed those of his
friends who still hoped for compromise. Henry had sent an artfully phrased letter to him at Montpellier, France. "If you can
recover in no other way," he wrote, "why not resign your seat
and leave public life for two years; five years; ten years, if
necessary, and devote your whole time to recovery . . . The
great question that you feel so much interest in, is not a question of a day. . . . If you will go and travel in Siberia, I will
leave German, Law, Latin, and all, and go with you, and take
care of you, and see that you don't speak a word of politics or
receive a letter or a newspaper for the next two years." Perhaps
alarmed at the evidence of Seward's rising influence upon the
Adamses, Sumner quickly reassured his young correspondent
that his health was much improved and that he expected soon
to resume his seat in the Senate. Writing again with even more
apologies for his presumption Henry tried to insinuate a doubt
of Sumner's physical readiness, but the artifice failed.[8]

What Henry envisaged for himself in the event of their
father's triumph one can only guess. In a nation in which Manifest Destiny had added a new dimension, even a boy could see
there was no plethora of qualified leaders. The history of his
own family suggested that no dream might be too fanciful. Had
not his grandfather been American minister to Russia during
the presidency of Henry's great-grandfather and in turn become president? But dreams were risky; out of dreams often

came crushing humiliation. Could their father lead them through the wilderness of passions that darkened the land? The assault upon Sumner and John Brown's raid signalled danger ahead; a few more affairs like these, "then adieu my country." With more foresight than he realized he advised Charles: "In America the man that can't guide had better sit still and look on. I recommend to you to look on, and if things don't change within a year then I'll eat my head." Things did change and the Republican party, again including Congressman Adams, swept to victory with Abraham Lincoln. But Henry did not foresee that the "comparatively conservative men" were destined to be pushed aside and that Senator Sumner who had helped the Adamses to rise in politics was, as a Radical Republican, to rejoice in their fall.

A Pleasant Series of Letters

In February 1860, a dramatic episode in the House of Representatives indirectly led to Henry Adams's first job as a newspaper correspondent. His father, who had been spending his first term in the House in dignified silence, disconcerted his colleagues in the Massachusetts delegation by announcing that he would not go along with them in approving the routine appointment of a spoilsman House printer. Through two weeks of party pressure and sixteen roll calls he stubbornly stood by Stanton of Ohio in his attack on political pork. Stirred by the "spectacle of an honest man in Congress," Henry praised his father's stand as "the first declaration of the colors we sail under." He had just come upon the story in the first of a series of articles in the Boston *Advertiser* above the signature "Pemberton." The author, he learned a few days later, was his brother Charles, who had written the article as a trial balloon to test public response to his father's stand. The subterfuge delighted Henry, for the unsuspecting Boston *Courier* promptly came to the defense of the Congressman.

The Pemberton letters "stirred" him up on the eve of an extended tour of Italy. "It has occurred to me," he wrote Charles, "that this trip may perhaps furnish material for a pleasant series of letters . . . If you like the letters and think it would be in my interest to print them, I'm all ready. In any case you can do just what you choose with them so long as you stick by your own judgment. But if, under the absurd idea that I wish to print, you dodge the responsibility of a decision, and a possible hurting of my feelings; by showing me up to the public amusement without any guarantee against my making a slump, you'll make a very great mistake." Charles, who did not share Henry's almost morbid diffidence, readily accepted the responsibility. As a result the first of the "pleasant series of letters" appeared in the proslavery Boston *Daily Courier* on Monday morning, April 30, 1860. It must have been a disappointment to Henry that Charles had not been able to find a more impressive medium than "the respectable old *Courier* and its two hundred supporters." The appearance of a private correspondence was soon discarded and the last four of the six letters were printed simply as "Correspondence of the *Boston Courier*" under such captions as "Letter from Austria," "Letter from Italy," "Letter from a Tourist," and uniformly signed "H.B.A.," a disguise that none of Henry's friends could have had difficulty in penetrating.

Light as his purpose professed to be it would have to be carried out against a background of uncommonly serious events, for resurgent nationalism was convulsing Europe. The War of Italian Liberation had come in 1859 but the opportunist Napoleon patched up the shabby treaty of Villafranca just in time to betray the hopes of the Italian patriots for a liberated Venetia. More was lost than Venetia, for in 1860 Napoleon extorted

Savoy and Nice and allowed Cavour to rescue the Italian monarchy from the socialistic evils of Mazzini's republic. Adams had caught a glimpse of the New Italy on his earlier visit in August 1859. Now he was back in northern Italy witnessing the excitement attending the formal annexation of Parma, Modena, Tuscany, and Romagna. There he heard the fabulous news of Garibaldi's landing at Marsala on May 11. A motley brigade of poorly armed Italians with a sprinkling of English and American sympathizers had successfully carried the assault. At the moment the expedition was to sail, Cavour's government had withheld the modern Enfields bought by the American and British committees. The equivocal behavior of the government did not deter Garibaldi and he and his thousand went on to Palermo where they outwitted the garrison of more than twenty thousand Bourbon and mercenary troops and received the capitulation of the Toledo on May 30, 1860.

Since Adams aimed at what is now called "human interest" stories, he eagerly sought out the picturesque and the dramatic, even the melodramatic. As he said the assault on Palermo was only "the first act of the melodrama." He deliberately ignored, therefore, the purely military and political facts reported by such regular correspondents as Henry Raymond of the New York *Times.*

All manner of persons caught his roving eye and each he characterized with engaging and often self-revealing candor. A fortune-hunting lieutenant aboard the Vienna express diverted him with foppish "prattle" and started a train of uncomplimentary reflections on the Austrian intellect. When he caught sight of the carriage of Emporer Francis Joseph he set out in hot pursuit. "So long as I have good health and am no misanthrope," he told his readers, "I mean to satisfy as far as I can a healthy and harmless curiosity." King Victor Emmanuel II disappointed him, for His Majesty "looked like a very vulgar and coarse fancy man, a prize fighter, or horse jockey." At the opera he scrutinized the imperturbable Cavour through a pair of binoculars, "a most quiet, respectable looking, middle-aged gentleman. From his appearance I never could have guessed that he was the greatest man in Europe."

The most important figure in his gallery of vignettes was, of course, that of Garibaldi, the Dictator of Sicily, but his touch was still fumbling. It would take years to hammer out the appropriate epithet. His friends having successfully pulled wires for him, he reached Palermo about a week after its fall. Mounting the grand staircase of the captured palace where "one saw everywhere the headquarters of revolution, pure and simple," he pushed upward through the disorderly crowd until, suddenly, "there we were in the presence of a hero . . . He had his plain red shirt on, precisely like a fireman, and no mark of authority. His manner is, as you know of course, very kind and off-hand, without being vulgar or demagogic. . . ." It was the fashion of Europeans, said Adams, to call Garibaldi "the Washington of Italy, principally because they know nothing about Washington. Catch Washington invading a foreign kingdom on his own hook, in a fireman's shirt! You might as well call Tom Sayers, Sir Charles Grandison."

Occasionally a note of satire flickered for a moment and then went out, casting the chaste image of the Genteel Tradition. In Rome, he discovered that "everyone seems to have a rage after Venuses, from painted ones to fettered ones, and yet it is tolerably safe to say that a statue of Venus, especially a nude, in one's parlor is bad taste, and still more, that, usually, a Venus is the most insipid and meaningless work an artist ever makes." Crudity of another sort was the mark of a certain type of American politician. "One of our gentlemanly Democratic ministers abroad, once crossing a frontier in his normal condition of crazy inebriation, refused to show his passport," at last flinging it into the officer's face. "The principle was correct," commented Adams, "but the manner faultily suave." It was the same sort of crudity that prevailed at home. "There is a general sensation or suggestion of bad Bourbon whiskey about American politics that is not pleasant," he complained. "Art must exercise a refining influence, and a man who comes here to pass his life, drops Bourbon whiskey and takes to lemonade or Bordeaux at least. Americans who live abroad read the American papers with a sort of groan."

These Italian letters highlight the Brahmin cast of his idealism. He tried hard, as he promised at the beginning of the series, to be "fair and unprejudiced," but when he saw the howling Sicilian mobs he instinctively flinched. "Where is the Sicilian nobility and the gentlemen who ought to take a lead in a movement like this?" he broke out. "One cannot always control his ideas and prejudices. I can never forget, in thinking of Sicily and the Kingdom of Naples, that under the Roman government these countries were the great slave provinces of the empire, and there seems to be a taint of degradation in the people ever since. It is not good stock." Similarly, the proposed Sicilian plebiscite seemed to him a travesty on democracy. He confessed that "these European popular elections have a little too much demonstration in them; they are a sort of continental squatter sovereignty, and very like a satire on our theories."

The important thing for Americans was to "reserve judgment until both sides have been heard," for, as he put it in another dispatch, "It is always better and pleasanter to look at more than one side of a question." As other correspondents had allowed themselves to be carried away by emotional partisanship Henry Adams labored to find a middle ground of truth. "The Austrian government," he acknowledged, "is a mass of faults and evils, but even republics are not always wholly pure." He was much more skeptical of the charges of tyranny levelled against the Grand Duke of Tuscany. "I rather fancy that the government was on the whole an exceptionally good one, and the people among the most happy and honest in all Europe."[1]

In Rome he likewise sought to redress the balance. Of a general who had been bitterly criticized for his bloody dragonnades in the service of the Papal government, he wrote: "A man more copiously abused than this good General, I have seldom heard of. Of course all the liberals hate him with a really cordial hatred." The most striking example of Adams's impartiality was his defense of the recently dead Bourbon, Ferdinand II. "So about the King, I feel more pity than pleasure at his troubles . . . It is the fashion to abuse him, just as it is the fashion to abuse the Pope and the Grand Duke of Tuscany; but you would probably find that these are all good men enough, just as good and very likely a great deal better than you or I, or the writer of the London *Times,* who tears a passion to rags splendidly."

If these opinions seem painfully superficial, we should remember that they are the complacent insights of a very young man, for Henry Adams had just turned twenty-one. Besides, it was almost inevitable that one of his name should gravitate toward the well born and conservative society of Vienna and Rome, a society to which he had almost official entrée. In those circles he was not likely to hear an unvarnished tale of the popular grievances. The devout *signori* of Rome, among whom the staff of the American legation freely moved, would not be likely to tell him, for example, that one of the Pope's recent medals commemorated the work of Colonel Schmidt at Perugia, who, in putting down the popular uprising, sacked the city and massacred its defenders. There was no one to remind young Adams, as he strove to be fair to both sides, that the best witnesses against the lay and ecclesiastical rulers could only be interviewed in the prisons and cemeteries of Central and Southern Europe. There was no acquaintance to point out to him that the Italian republic had been destroyed by the pressure of the very "commercial interests" in England, France, and America with which he was accustomed to identify himself and his family.[2]

Adams, as he had been frank to admit, had his own share of

prejudices. Being a member of a society dedicated to abstract moral principles and provincial gentility, he could not without violence to his heritage identify himself with the desperate aims of the European masses, nor could he reconcile those aims with his belief in the inevitable—and gradual—progress of mankind, progress under the leadership of Guizot's "rich and enlightened classes." [3] His passion for order and propriety betrayed him into making excuses for the bad old times and the expensive amenities of continental court circles. As a disciple of Edmund Burke he inclined to pity the plumage and forget the dying bird. In short, as one charts the drift of these letters one sees that the nearly two years which Adams spent in Europe did not change his outlook. He was ready to return to America, the "Boston standard of thought and the mental scale of Harvard" approved by experience.

Upon the publication of the last dispatch, George Lunt, the editor of the Boston *Courier* and one of most "stiff and ultra of the conservative determination," [4] appreciatively commented: "The letter from H. B. C. [*sic*] is the last we shall have from him on Italian affairs. Our readers, who have been highly entertained will regret this; for the letters have been much above the average of communications sent home by European travelers . . . What he has gathered abroad he will be sure to turn to advantage at home." Whether Adams received anything more substantial by way of payment is not now to be learned. At least he had the satisfaction of knowing that he had struck precisely the right note for the newspaper. In any case both he and his brother Charles could now feel he had taken a first step toward success.

The time had in fact come for Henry Adams to return home, whether it was sure or not that he would be able to "turn to advantage" his European experience. Drawing to a close was the two-year holiday agreed to by his father and yet he had come no closer to the choice of a career. Now, without leisure for debate, he had to make up his mind. The ordeal plunged him into gloom once more. He felt himself "getting old and cautious." Full of disillusionment at twenty-two, he wrote to his mother from Paris: "I find that people are unhappy everywhere and happy everywhere." Charles broke the spell by urging that Henry should join his parents in Washington and study law there, as their father's re-election was certain. Relieved in mind, Henry clutched the proposal and made his submission. "I shall make up my bed in Washington, and no doubt it will be just as pleasant as anywhere else," he reassured his mother. "At all events, whether it is or not, it's the place that my education has fitted me best for, and where I could be of most use. So if Papa and you approve this course, and it's found easy to carry it out, you can have at least one of your sons always with you." Thus literature and the quiet literary life were set aside and, misgivings and all, the dutiful young man betrothed himself to the Law.

Chapter Three

Washington Correspondent
1860–1861

Disciples of the Massachusetts School

W HEN HENRY ADAMS reached Quincy in October of 1860, shortly before the November sixth election, the domestic situation, which had looked almost Utopian set against the chaos of Italy, no longer justified any feelings of complacency. North and South once again jockeyed for control of the national government, but a deeper note of urgency now entered the contest. Leading South Carolinians "unanimously resolved that in case of Lincoln's election the state must secede." Other Southern firebrands threatened to prevent Lincoln's inauguration by violence if he should be elected. Fairly accurate rumor placed sixty-five thousand men under arms in South Carolina alone. Even in Quincy the "air" of the young "Wide Awake" Republicans who paraded in disciplined columns through the streets left Adams with an impression that "was not that of innocence." [1]

To the white South the impending election seemed the last political redoubt guarding its slave-based way of life. Southern statesmen had begun to see the shape of things to come in 1856 when the new Republican party agreed in convention to establish a frankly Northern party dedicated to resist the extension of slavery in the territories. As if to accept the challenge, the Supreme Court in the Dred Scott case voided the Missouri Compromise prohibiting slavery north of 36°30'. According to Chief Justice Taney a Negro had in effect no rights which a court was bound to respect. Forty per cent of the human population of the Cotton States were thus guaranteed the legal and political status of domestic animals. The barrenness of that legal victory for the firebrand view of the Constitution soon became clear. What happened in the territory of Kansas in the defeat of the proslavery Lecompton constitution foreshadowed the ultimate recession of the political power of the South. It became increasingly clear that a Republican victory, stripping the South of its political bargaining power, would establish that section as an economic satrapy of the Northern manufacturing states, helpless to resist encroachments on the peculiar institution to which its wasteful economy was wedded.

Fresh from the Paris of Napoleon III, young Adams was about to be projected into this final desperate struggle for power, a struggle which before it had run its course through Civil War and Reconstruction would severely test the political wisdom of the Adamses. Three generations of family statesmen had groomed him for the conflict. No matter what the name of the party that transiently embraced their view of the moral and political order—whether Federalist, Whig, Free Soil, or Republican—the family had supported its cause as sacred only so long as the party held to the Adams creed. John Adams had insisted on his moral right to govern free from the dictation of such party leaders as Alexander Hamilton and the effort cost him re-election. His son John Quincy Adams had flouted party discipline to follow his "sense of duty" in supporting President Jefferson on the Embargo Question. This act permanently alienated him from the Federalist party. When the Mexican War threatened, Charles Francis Adams exercised the moral prerogative bequeathed by his father and grandfather and helped lead the "Conscience" Whigs out of the party.

Long before Seward's "Higher Law" John Quincy Adams had enunciated the doctrine that "the eternal and immutable

laws of justice and morality are paramount to all human legislation." In season and out, in public declaration, private letter
and diary, the members of the family asserted those laws. It
was John Adams who first taught them that it is a "false and
immoral maxim, that the end will sometimes justify bad means."
Charles Francis Adams likewise rejected that maxim as "the
delusive track of expediency." In a letter to his son Charles, in
which he held up John Adams as the exemplar of the moral
statesman, the elder Adams summarized the family's principles
in these words: "The first and greatest qualification of a statesman in my estimation, is the mastery of the whole theory of
morals which makes the foundation of all human society: The
great and everlasting question of the right and wrong of every
act whether of individual men or of collective bodies. The next
is the application of the knowledge thus gained to the events
of his time in a continuous and systematic way . . . The fee-
bleness of perception and the deliberate abandonment of moral
principles in action are the two prevailing characteristics of
public men . . . No person can ever be a thorough partisan
for a long period without sacrifice of his moral identity. The
skill consists in knowing exactly where to draw the line." [2]

The instruction in that skill was best taught in the "Massachusetts School" of statesmanship. For, as de Tocqueville had
demonstrated in the pages of *Democracy in America*, political
morality had its geography. "The civilization of New England,"
he wrote, "has been like a beacon lit upon a hill, which, after
it has diffused its warmth immediately around it, also tinges
the distant horizon with its glow . . . The emigrants of New
England brought with them the best elements of order and
morality" and created "a democracy, more perfect than antiquity had dared to dream of. . . ." This was the salutary moral
order that Henry Adams believed would eventually have to be
implanted in the South if it were ever to be redeemed from the
"Slave Power." It would "depend on the generation to which

you and I belong," he once lectured his brother, "whether the country is to be brought back to its true course and the New England element is to carry the victory." To their ex-President grandfather the existence of slavery was "incompatible with the Law of Nature and of Nature's God, which has given to all men the inalienable Right to Liberty." [3]

To Henry Adams and his forebears the compromise with slavery had been the one blot upon the Constitution. It became their lifelong occupation to erase the blot without destroying the fabric, for it was their faith in the Constitution, "this side idolatry," that sustained them through generations of political adversity. That document according to John Adams embodied a system of checks and balances that reproduced in human society the "triple balance" exhibited in the constitution of nature itself. By the appointment of Chief Justice John Marshall he insured the faithful maintenance of that equilibrium. [4]

In view of this tradition, Henry's father, though a "determined anti-slavery man," had no sympathy for the Abolitionist Garrison whose *Liberator* slogan denounced the Constitution as "a covenant with death and an agreement with hell." There were other Union men who saw eye to eye with Charles Francis Adams, men like Richard Henry Dana, John G. Palfrey, Dr. S. G. Howe, and Charles Sumner, and these men met at the Adams home on Mount Vernon Street under the inquisitive eye of the boy Henry. From them he received his earliest lessons in the art of moulding public opinion when they dedicated the Boston *Whig* in 1846 under his father's editorship to saving the Union through the "total abolition of slavery." One of Dana's letters to Henry's father throws a light upon the confused aspirations of this middle-of-the-road group. "If the Union can be preserved with an unmutilated Constitution, unclogged by new compromises, I am willing to support a reasonable Fugitive Slave Law; for I regard the preservation of such a Union as a very high moral duty, which we owe to peace, to civiliza-

tion, to the development of a continent, and to the founding of
free states, justifying the most painful legal duty of rendering
back a fugitive slave." [5]

It was on this rock that the family foundered. Its members
tried to unite in a single moral imperative the twin duties of
maintaining the Constitution and of opposing slavery. The effort to reconcile what was hopelessly irreconcilable finally
drove them out of the Republican party when the extremists,
uniting conscience with fanaticism, captured control of the
party.

Whatever moderation Charles Sumner had felt vanished on
the day Brooks struck him. As the war came closer, Sumner's
extravagances began to embarrass the Adamses, with whom he
was still on the friendliest terms, dining regularly with them
each Sunday as he had for nearly a dozen years past. His vulgar ranting attack in the Senate chamber on the "Barbarism of
Slavery" in June 1860, contrasted disagreeably with the moderation and high-minded dignity of Charles Francis Adams's
maiden speech in the House a few days before.

Adams defended the Republican party as a necessary coun-
terpoise to the "Slave Power," weighted with the noblest of
missions. "The cardinal principle of the Revolution," he said,
"that which marks a real advance in the progress of political institutions, is that the individual man, whether in or out of the
social organization, whilst doing no wrong, has certain rights
which his fellow-man all over the globe is bound to respect.
The general establishment of this maxim in the practice of the
nations of the world is, or ought to be, the mission of America
during the present century of her growth." The Boston *Traveler*
fatefully commented, "Those who were disaffected by Mr.
Sumner's speech cannot fail to admit the power of that of Mr.
Adams." It was the first hint of the rivalry that had already
begun between the two old friends.[6]

The political orientation of the family had steadily shifted toward center as Seward of New York and his astute manager, Thurlow Weed, rose to leadership in the Republican party. Henry's father chose to follow Seward's "footsteps" because, as he later said, Seward "alone, of all others, had most marked himself as a disciple of the school in which I had been bred myself." Charles Francis Adams had gone to the Chicago convention convinced that Seward would be chosen as the Republican nominee for President. Under his leadership eighteen of the twenty-six delegates from Massachusetts held firm for Seward on the third and deciding ballot, but the defeat of Seward was decisive. To Adams's chagrin, Abraham Lincoln received the nomination. In the interest of party solidarity, Seward swallowed his disappointment—and at once began his efforts to slough off the radical elements in the party. Recognizing the value of his Massachusetts ally as an offset to the intractable Sumner, Seward visited Quincy and persuaded Charles Francis Adams to join him in a swing through the then-Northwest in behalf of Lincoln's candidacy. Henry's brother and confidant, Charles Francis, Jr., accompanied the party and himself took the stump.[7]

At this juncture Henry returned from Europe. Loyal to his party, he voted for Abraham Lincoln on November 6. At the same election his father, standing for re-election, defeated the Union party candidate for Congress by a very comfortable majority. Election day was also memorable to young Adams as marking the formal beginning of his study of law in the office of the influential Judge Horace Gray. But there was to be as little progress in the Common Law as there had been in the Civil Law. Three weeks later he was in Washington as private secretary to his father in time for the opening on December 3 of the tumultuous "lame duck" session of the Thirty-sixth Congress.[8] Before leaving Boston he managed to make connections with "the most important" Republican paper in Boston, the *Advertiser,* whose co-editor, Charles Hale, a good friend of the family, appointed him Washington correspondent. Expediency required the appointment to be kept secret.

In the national capital the private secretary and newspaper correspondent immediately displaced the half-hearted law student. The first letter sent to Charles from Washington a week after his arrival indicated where his interests now lay. "It's a great life," the young statesman-in-waiting exulted, "just what I wanted." There was no time for the quiet study of Blackstone's *Commentaries* in the throbbing atmosphere of the capital where crisis succeeded crisis and treason seemed to lurk behind every window blind. "What with the duties of secretary, of schoolmaster, of reporter for the papers, and of societyman," Henry explained, "I have more than I can do well." So far as the remainder of the winter went, law was sunk without a trace not to reappear again until the following March in Boston and then only briefly.

From the beginning the young secretary realized that he was an eyewitness to history. He therefore announced to his brother that he would record his testimony for posterity. "I propose to write you this winter a series of private letters to show how things look. I fairly confess that I want to have a record of this winter on file, and though I have no ambition or hope to become a Horace Walpole, I still would like to think that a century or two hence when everything else about it is forgotten, my letters might still be read and quoted as a memorial of manners and habits at the time of the great secession of 1860."

His work as Washington correspondent of the Boston *Advertiser* once under way he extended his schoolmastering to fellow members of the press. Even in the cloakrooms of Congress he dispensed "good Republican doctrine and lots of it." His source of information was indeed the fountainhead of such doctrine. Republican leaders like Seward, Sumner, the distinguished apostate "Know Nothing" Henry Winter Davis, and a crowd of lesser politicos of all shades of resolution made their rendezvous in the Adams parlor. Usually Henry sat against the wall at these councils, hastily transferring his incisive observations to his "memorial of manners and habits" for Charles's eye. Once, however, he "squelched" the headstrong Sumner by reading a few lines from Bacon's essay "On Seditions and Troubles" while Sumner was busy parrying a violent rebuke from Charles Francis Adams, the new "Archbishop of Anti-Slavery" who was beginning to wear the miter of authority.

Drawn into the vortex of behind-the-scenes negotiation, Henry experienced a sense of "continual intoxication." It was "magnificent to feel strong and quiet in all this row, and see one's own path clear through all the chaos." He promised Charles that his reports to the *Advertiser* would be on "the crescendo principle," and ordered "if the battle 'should wax hot,' and Charles Hale does not rise to it, you must thumbscrew him a little."

If he gave advice he also willingly received it. To Charles, who was as usual unsparing in his criticism, Henry admitted: "Naturally it is hard at first for a beginner as I am to strike the key note; still I think I can manage it in time; and meanwhile criticize away just as you please." On occasion he took issue with his brother, as when he feared that his cloak of anonymity might be snatched off. "You caution boldness," he charged, "at the very time when a bold slip might close my mouth permanently. It was but this morning that C.F.A. [Charles Francis Adams] cautioned me against writing too freely." It was essential that the identity of "Our Own Correspondent" should remain secret if he was to maintain a foot in both camps, the moderates and the "Ultras." The New York *Times* made him nervous by reprinting one of his letters "with copious italics," and he feared that his disguise was penetrated. "Can they suspect or have they been told whence they come?" His fears proved groundless, there being no hint in his later correspondence that his father's opponents ever found him out.

In the Field of Grain

In his initial dispatch to the *Advertiser*, dated December 7, 1860, Henry Adams reported that so far as he could learn "the feeling here is that nothing will do any good until secession has been tried." Doubtless he shared the hope, recorded by Charles in his journal, that secession would be tried, "for the country is weary of the threat." Charles faced the "irrepressible conflict" with equanimity. It must be abortive as it would "cost the States which try it out about ten millions of dollars" and, as his father had argued in the House, would "fail ignominiously." Henry expressed this point of view in the following two contributions; but when the Seward strategy entered the final, delaying phase of the political struggle, designed to span the dangerous interval preceding Lincoln's inauguration, Henry promptly abandoned the line of self-confident indifference for that of honorable conciliation. His professional task thereafter required that he persuade his father's constituents that "conciliation" did not really mean "compromise" or the surrender of principle.[1]

At the outset, in order to prevent any overt acts that might upset his father's subtle maneuvers in the Committee of Thirty-three, he sedulously played down the bitter wranglings in and out of Congress. The second letter, for example, reassured his father's constituents that "It would be an insult to the great leaders of the party to suppose that their ideas on this matter are changed; that they are political gamblers, always on the lookout to hedge or bluff as the cards turn up. Their theory has been declared over and over again that disunion is an impossibility. A mere temporary secession is not disunion nor anything approaching it."

Two weeks later, nonetheless, he was praising the "last re-sort" proposal of H. Winter Davis of Maryland, "the great leading spirit of the middle men," who proposed the admission of New Mexico as a sop to the border states. To his brother he wrote with much less confidence: "You will see by my *Advertiser* letter our ideas about compromise, and you will understand that we would yield a good deal to avoid a split now which would be very bad. C. F. A. is decided to vote for Winter Davis' proposition, but this is private. It may never come up." By January 1 he was obliged to defend his father against the charge of "backing down," an unpalatable accusation made by the Boston *Atlas and Bee* with its usual venom.

The problem that faced Henry's father and his chief, Seward, of maintaining party unity without relinquishing the slippery reins of party leadership to the militant radicals was one of extreme delicacy. The "Ultras" did not hesitate to interpret Charles Francis Adams's statements in the worst light possible. A few of Sumner's correspondents thought that Adams deserved to be classed with Benedict Arnold. On the other hand, the right-wing Republicans in Massachusetts, whose petitions and memorials urging conciliation upon Congress bore over thirty-five thousand signatures, must have been equally annoyed at the meagerness of Adams's proffered concessions. Yet the appearance of compromise needed to be shown if the Republican party was not to follow the Radical wing into the trap set by the Southern strategists. At all costs the onus of rejecting conciliation would have to be fastened upon the Southern representatives or else the Republican party would be accused of treasonably desiring disunion. Charles Francis Adams hoped that the bargaining could be prolonged at least until the Republicans should take over the Administration on March 4. "Our only course in the defenseless condition in which we found ourselves," he privately acknowledged in April, "was to gain time, and bridge over the chasm made by Mr. Buchanan's weakness." [2]

With events moving in kaleidoscopic fashion, with factions North and South clamorously jockeying for position and all working at cross purposes, it was no wonder that the young correspondent had difficulty striking the right keynote as demanded by Charles. The task would surely have bewildered a veteran publicist. What impresses the present-day reader is the aplomb with which Henry at twenty-three skillfully sailed the course plotted by Seward and his father, contriving somehow to tack and run before the wind almost simultaneously. The curious alternations in that policy caused Oliver Wendell Holmes to grumble disgustedly, "If Mr. Seward or Mr. Adams moves in favor of compromise, the whole Republican party sways like a field of grain, before the breath of either of them. If Mr. Lincoln says he shall execute the laws and collect the revenue, though the heavens cave in, the backs of the Republicans stiffen again." [3]

During January the confused talk of coercion by the Radicals brought the country face to face with the possibility of war. Public opinion scrambled back in horror and many New Englanders seemed ready to accept even the most craven compromise. Obviously the position of the Republican party needed clarification if it was not to disintegrate into the medley of factions out of which it was created. Charles Francis Adams rose in the House on January 31 to define a reasonable middle ground. Reviewing the events in the South, he professed to see no evidence of a preconceived plan of disunion except of course in South Carolina. There the treason was deliberate. He shrewdly rehearsed the stated grievances of the South, attacking in each case the fallacy on which he believed it was grounded, then placed the blame for aggression upon the South. Nevertheless he rejected the idea of preventing secession by military means. Another course remained open. "I see no obstacle," he said, "to the regular continuance of the government in not less than twenty states, and perhaps more." [4]

The solemn appeal by the acknowledged spokesman of the Sewardites carried great weight; and it seemed for a while that the nation had passed its dangerous corner. But Sumner's absence on the day of the speech was duly noted. Rumors quickly circulated that the two great rivals had quarreled. Henry, hoping to prevent a disastrous rupture between the two men which might split the party, glossed over the matter in the columns of the *Advertiser* in an effort to flatter Sumner into silence. "By the way," he wrote with studied casualness, "I see various rumors about a quarrel . . . [the] stories are very unfair to Mr. Sumner indeed, and are only one more example of the evils of 'sensation reports.' There has been no quarrel between these two gentlemen." He was considerably more frank in his report to Charles. "As for Sumner, the utmost that can be expected is to keep him silent. To bring him round is impossible. God Almighty couldn't do it. He has not made his appearance here for more than a week, though there is as yet, so far as I know, no further change in the position of matters between him and C. F. A. As usual I suppose he will stand on his damned dignity . . . We shall do all we can to prevent his bolting." He suspected however that the "whole Garrison wing are doing their best to widen the breach."

In truth the breach could not possibly be kept from widening. The antislavery element in the party was hopelessly divided between conservative and radical, a fateful division that was to plague the whole course of reconstruction during and after the war. Especially bitter was the tug of war over the Cabinet appointments, which broke out shortly after Adams's speech. As a politician Senator Charles Sumner may have conceded the expediency of Lincoln's choice of Seward for the State Department, although he did not give up hope that Seward might somehow be either kept out of the Cabinet or ultimately driven from it. But he was hardly prepared for the indignities which followed. Seward, ignoring Sumner's wishes, sent Thurlow Weed to Springfield to urge the appointment of Charles Francis Adams to the Treasury. Lincoln prudently rejected the suggestion, but he also rebuffed Sumner's nominee, Salmon P. Chase. Seward then "begged very hard" for the English mission, and the President-elect acquiesced because "really, Seward had asked for so little." [5] That "little," however, was deeply coveted by Sumner whose friends had worked hard to prevent the appointment of Adams. It was particularly galling that the man whom he had helped raise to office should thus become the instrument of his humiliation. He did not need the comment of the New York *Herald* to know that Seward had advanced Adams in order to get even with him and his radical following for their hostility. Henceforward, he would inflexibly oppose the political ambitions of the Adamses.

The remaining four *Advertiser* letters, which followed the speech of Charles Francis Adams, continued to reflect the ebb and flow of the hopes entertained by the moderate Republicans. Henry, who had been delighted by his father's presentation of their case, supported the wisdom of Seward's pacificatory efforts. Thus when Seward expressed approval of the conciliatory action of the Virginia legislature in proposing a Peace Conference, Henry publicly backed him up. "For my part, I prefer 'the pleasant optimism' of Governor Seward, who hopes that this step will so strengthen Union men in the border states that a month hence secession will be one of the nightmares of the past." His last dispatch continued to hew to the line: on the one hand he loaded reproaches on the uncompromising "Ultras" who proposed disunion or coercion as the only solutions, on the other, he reprimanded the disloyalty of the Massachusetts citizens who would appease the South with the Crittenden compromise.

The series of contributions ended when the editor, Charles Hale, himself came to Washington to carry on as Washington correspondent. Young Adams, heartily "sick" of the "hard work" entailed by "this temporizing policy," was quite ready to quit, although Hale had been "complimentary in his remarks generally." The contributions had cost him many late hours at his desk, and one suspects the political virtue of his effort was its sole reward. Chiefly he had been annoyed by Hale's editorial scruples which had time and again sacrificed Adams's choicest libels against the objects of his contempt. Sometimes the censorship went so far as to suppress the whole of a dispatch.[6]

Though shorn of their "spicy parts," the *Advertiser* contributions gave proof of Adams's growing literary powers. There was a surer touch in the selection of details and an interest in political figures beyond the requirements of mere partisan rhetoric. Likewise evident was his deepening sense of the dramatic. The characterizations varied from a glancing hit at Burnett of Kentucky "who has a voice like a bull and a face not very unlike his voice" to a set piece on the self-acknowledged disciple of John Quincy Adams, Henry Winter Davis of Maryland, a piece in which the Adamsian consciousness of history had full play. "When the history of our great secession comes to be written, a century or two hence, the historian ought to make a parenthesis to describe Mr. Winter Davis. He ought to describe him as he rose in his seat today, and stood with his arms folded, in rather a studied attitude, waiting for the House to come to order. The democrats call Mr. Davis a demagogue, but no man ever had less the appearance of a demagogue than he. Rather short than tall; with a graceful figure; a finely-cut expressive face; crispy hair, close-cut so as to show a finely-shaped head to the best advantage; remarkably neat and well-dressed in his round-cut English clothes, Mr. Davis gives one the idea of rather an aristocratic person, and forms a very striking contrast to most of his associates."

The relationship with the *Advertiser* was unsatisfactory on at least one other ground—its limited circulation. Desiring more scope for his efforts and wishing perhaps to counteract the pro-Sumner line of the columnist "Warrington," he asked his brother to arrange "to have the columns of the Springfield *Republican* open" to him as well. Editor Sam Bowles, though one of their father's most loyal supporters, did not accept the proffered services.

The *Advertiser* articles did produce a discernible effect upon editorial opinion. Much to Henry's relief, Charles Hale, who had inclined somewhat toward extreme measures, finally brought his policy into line with that of Seward and Adams. The New York *Times,* as we have already seen, had shown "particular respect" toward his letters. On looking back over the winter's journalism, he was justified in believing that he had "done some good in sustaining papa at home" by "shaping the course of opinion in Boston."

The Great Secession Winter of 1860-61

From the time of his arrival in Washington late in November to mid-February when he resigned his job as correspondent, Henry Adams found little leiṣure for anything but the most urgent work. His father's highly literate constituents did not hesitate to make their wishes known. Remonstrances from "Garrisonians or men without weight" might be offset by the flood of letters praising his father's great speech, but for all, acknowledgments had to be written. In addition, he made transcripts of his father's statements for the use of the printer and the gentlemen of the press and kept in communication with party leaders in Boston. Long letters went, for instance, to Richard Hildreth, the historian and political journalist, aimed at holding him to his father's course. As Washington correspondent he also was obliged to cultivate his notable acquaintances "to hunt secrets" for himself and his brother Charles. In the course of these political and social expeditions he came to know one man who was soon to be unexpectedly useful to him, Henry J. Raymond, editor of the New York *Times*, and he had at least a fleeting first encounter with John Hay, who arrived in Washington as one of Lincoln's secretaries.

Back in Boston, Charles Francis, Jr.'s law practice was not very exacting, although in their father's absence he and Henry's eldest brother, John Quincy Adams, had some occupation in looking after the family's sizable financial and real estate holdings. In the middle of the winter Charles resumed his prodding of Henry, suggesting a substantial essay on the current political situation. Henry's first impulse was to back away, but on second thought he changed his mind, perhaps stimulated to do so once more by his brother's example. Charles began a series of letters in the *Transcript* to counteract the war talk which had been revived, perhaps inadvertently, by Lincoln's Indianapolis speech en route to the Capital. Over the signature "Conciliator" he approved the rumored surrender of Fort Sumter as "a wise, a statesmanlike and judicious move." [1] At the same time he was hard at work on an article for the April *Atlantic Monthly* called "The Reign of King Cotton," in which he predicted that that reign would come to an end, sooner or later, simply as a result of internal strain. He based his prediction on the decline of popular education in the South. Soon other republican institutions would wither and chaos would follow. Only if the cotton monopoly were bankrupted and a new agricultural and industrial order gradually instituted could the South be rescued from despotism. All that needed to be done was to develop a cotton supply overseas and peacefully destroy the slave economy through competition.

With their father's appointment to the Cabinet or some other public post still hanging fire, Henry determined to write an article for the *Atlantic Monthly* entitled "The Great Secession

Winter of 1860-61." What he proposed to himself was to set the events of the preceding winter into a suitable perspective. The result was an unofficial White Paper defending the devious maneuvers of the moderate Republicans. Adopting Charles's attractive hypothesis, he analyzed Seward's strategy as designed "to prevent a separation in order to keep the slave power more effectually under control, until its power for harm should be gradually exhausted, and its whole fabric gently and peacefully sapped away."

He traced the current troubles to a source that had plagued the private reflections of his family since the year of the Missouri Compromise. "By an unfortunate necessity which has grown with its growth, the country contained in itself, as its foundation, the seeds of its future troubles. By the Constitution a great political, social, and geographical or sectional power within the Government was created; in its nature a monopoly; in its theory contrary to and subversive of the whole spirit of Republican institutions." The election of 1860 had challenged that entrenched power. Adams eloquently summarized the consequences of that challenge: "After a long and bitter contest the slave power was for the first time defeated, and deprived, not of its legitimate power, not of its privileges as originally granted under the Constitution, but of the control of the Government; and suddenly in the fury of its unbridled license, it raised its hand to destroy that Government . . . The great secession winter of 1860-61 was therefore the first crucial test of our political system." In his mind there could be no doubt that the system had stood the test. It might have met the test peacefully, he believed, if the Seward-Adams policy had prevailed.

Critically weighing the merits of the article, he perceived that he had not adequately solved the problem of form. The same feeling of disgust possessed him which, two years earlier, had caused him to bury his "Two Letters on a Prussian Gymnasium." Toward the end of April 1861, he made a gift of the article to Charles with the following note: "As you will see on reading it over, it is not worth printing. If it had been I should have given it to you before. But finding that it was not going to be a success, I just finished it and laid it by, thinking that though as a whole it is a failure, there are still parts of it which might be put to use." Charles acquiesced in his judgment. In a sense Henry Adams was making progress as an author. He now had two unpublished manuscripts.

He returned to Boston shortly before the middle of March for another go at Blackstone in the office of Horace Gray. A few days later on March 19, came the telegram announcing his father's appointment as Minister to England. Charles's diary vividly recalls the scene. "It fell on our breakfast-table like a veritable bombshell, scattering confusion and dismay. It had been much discussed in Washington, but Seward had encountered so much difficulty, and the President had seemed so intent on the nomination of Dayton, that the news finally came to us like a thunderbolt. My mother at once fell into tears and deep agitation; foreseeing all sorts of evil consequences, and absolutely refusing to be comforted; while my father looked dismayed. The younger members of the household were astonished and confounded." Charles's later comment on his mother's reaction suggests at least one source of Henry's pessimism. "My mother," Charles wrote, "took a constitutional and sincere pleasure in the forecast of evil." [3] Calm being restored, it was decided that Henry should continue his political apprenticeship as private secretary to the new minister.

While he marked time for their departure, an incident occurred which seemed to him to require his conciliatory services. John Gorham Palfrey, one of the old friends who had fought the Free Soil fight with his father, had just been appointed to the postmastership of Boston through the strenuous efforts of Charles Sumner. The appointment had been set afoot by

Henry's brother who hoped it would be the means of effecting a reconciliation with Sumner. Although Sumner came to a Sunday dinner at the Adamses, the amicable talk could not revive lost illusions. A few of the Boston papers resented the Palfrey appointment as the last in a distasteful series. For the sake of party harmony Henry prepared a letter for the New York *Times* in which he defended the appointments. Being also aware that the selection of his father for the English mission in place of Sumner had probably offended his boyhood idol, he thought to assuage that hurt also. To make sure that Sumner did not miss the item he sent a letter to him on the same day that he dispatched the contribution to the *Times*.[4]

The unsigned letter appeared on April 5 under the heading "The Federal Appointments for Massachusetts." Though undistinguished in style, the transparent rhetoric of the piece shows that the private secretary aimed to perfect himself in diplomacy.

In the first place Massachusetts has been much pleased at having the English Mission, and that it was given to Mr. Adams. The great body of people were indifferent whether it was given to Mr. Adams or to Mr. Sumner; either would have suited them, and either would have been excellent; all they wanted was that it should be offered to one of them, and since Mr. Adams was the one, they are perfectly satisfied . . . The last rub has been in the matter of the Post Office, which has caused a tremendous amount of swearing among the politicians, and has pleased the great mass of people very much. Mr. Sumner deserves very great credit for his appointment, not that he made it wholly on his own account, or without advice; on the contrary, he was supported by the strongest influences in the State, and did it only after long consideration, and the best advice . . .

Mr. Palfrey is one of the best and ablest men in the Repub-

lican ranks . . . It is a good act to draw him out of his forced quiet again . . .

The same thing almost exactly may be said of R. H. Dana, who is to be the District Attorney . . . Whatever the disappointed office-seekers may say, these nominations are the strongest that could have been made, and have given us great confidence in the Administration.

As further law study was now superfluous, Henry began to prepare himself for his new position. In the same letter to Sumner in which he referred to his *Times* letter he added that he had "a favor to beg" of him. "As I may have occasion in England to make some use of the press now and then, I want to know whether you can put me in the way, or can suggest a way, of getting an entrance into some of the English papers. As I shall be outside of the Legation and unconnected with it, I would like to act independently . . . I make this application, supposing that you in your time must have done very much the same thing and can advise me better than anyone else." Henry tactfully ignored the coolness between his irascible father and the equally irascible Sumner, but its chill fell upon his request.

By this time Henry's father had returned from Washington, having conferred with Secretary of State Seward and President Lincoln. What he was able to report of his interview confirmed all their fears of the President's incapacity. Lincoln listened "in silent abstraction" while Adams spoke his gratitude. The president replied that he could thank Seward for the appointment. "Then," as the *Life of Charles Francis Adams* recalls, "stretching out his legs before him, he said, with an air of great relief as he swung his long arms to his head:—'Well, governor [Seward], I've this morning decided that Chicago post-office appointment.'" The Minister who had hoped to discuss on the level of the highest statesmanship the course of American foreign policy that would need to be pursued during that critical period "never recovered from his astonishment, nor did the impression then made ever wholly fade from his mind." In the strict performance of duty Adams had never acted with regard to either praise or blame, but Lincoln's indifference was a startlingly new experience.

Although Charles Francis Adams was commissioned Envoy Extraordinary and Minister Plenipotentiary to England on March 20, he postponed his sailing for six weeks until his eldest son, John Quincy, was married to Fanny Crowninshield of Boston. The attack on Fort Sumter did not alter his plans, for it was unthinkable that the assault was anything more than a trifling insurrection and hence of little diplomatic concern. The Confederate envoys, however, had no equally pressing social obligations and got to London before him in time to inspire the Queen's Proclamation of Neutrality. Happily ignorant of the unpleasant diplomatic surprise that was impending, the new Minister and the private secretary set sail from Boston Harbor on May 1, officially with no more urgent business than the prospect of resuming the elegant court dress which had been proscribed by Secretary of State Marcy in 1853.

Chapter Four

London Correspondent
1861

Outside of the Legation

O NE of the influential persons with whom Henry Adams became acquainted during the secession winter in Washington was Henry J. Raymond, editor of the New York *Times.* They saw eye to eye on political realities. Raymond had vigorously supported Seward's candidacy in the Chicago convention and like Charles Francis Adams had thereafter thrown his weight behind the much-criticized program of conciliation. When the administration announced Charles Francis Adams's appointment as Minister to England Raymond editorialized: "There is no man in the United States more conspicuously fitted for that highest of the diplomatic posts in the gift of the government than he." Before young Adams left for England, arrangements were completed for him to act as London correspondent of the *Times.*

The new arrangement required even greater secrecy than the one with the *Advertiser.* As the State Department prohibited "all communications with the press," the new Minister was "very careful to impress upon all the members of the Legation the importance of obeying the injunctions." [1] No one knew that Henry was to be a London correspondent except his brother Charles, who later acted as his agent in collecting the quarterly payments from Raymond. It must have been easy enough to keep the secret from Raymond's subordinates, but more than ordinary dissimulation would be needed to keep Henry's father in the dark, especially as father and son spent whole days together writing opposite each other in the delightful old study of the Legation at 5 Mansfield Street, "as merry as grigs." Henry could have justified his conduct by arguing that as private secretary he was not actually employed by the State Department; still this technicality would hardly have been respected either by hostile critics or by his scrupulous father.

The London correspondent proposed to create a favorable public opinion in the United States in support of his father's diplomatic policies. Again, as in the *Advertiser* series, his problem was to allay embarrassing public clamor, this time the sort of clamor aroused by such viciously anti-British sheets as James Gordon Bennett's New York *Herald*. Again the job was one of utmost delicacy. Minister Adams defined his own mission as the reëstablishment of "confidence between the countries which has been somewhat shaken of late." Henry had ample reason to know that the responsibility for that shaking could be placed on certain leaders of the Republican party. Seward, for example, had suggested in 1860 that the loss of the Southern states might be offset by the annexation of Canada and Mexico. In his *Advertiser* letter of February 8, 1861, Henry had made a veiled allusion to a countermeasure proposed by the Radicals. A reporter for Bennett's *Herald* eagerly revealed the annexation scheme and he quite accurately identified it with Sumner. Edward Everett, one of Henry's uncles, thought the plan "a magnificent project." In England, on the other hand, the cotton processors and their allies played into the hands of warmongering factions in the United States by urging immediate diplomatic recognition of the Confederacy. These patriotic factions were angered by the Proclamation of Neutrality which had been issued while Minister Adams was en route to his post; they also envisaged war with England as a providential cure-all for domestic troubles.[2]

Seward's first instructions required that the Minister protest the issuance of the Proclamation of Neutrality and discourage further communication between the British Ministry and the Confederate envoys, a formidable task at that moment. It was made no easier by the attacks of "that old beast Gurowski," the journalistic Don Quixote of Washington, who construed the Minister's acquiescence in Seward's policies as the servile "obeisance of a corporal." [3]

Henry continued to collaborate with his brother, occasionally writing to him to use his influence with the Boston *Advertiser* —as he had done before—to procure editorial support for the line laid down in his own dispatches. He also attempted, though without success, to obtain a place for Charles as American correspondent for a London newspaper. In an effort to counteract the Southern propaganda he steadfastly worked "to get some influence over" the London press, but though he ocasionally "worried a newspaper writer" he advanced slowly. The hoped-for introductions from Charles Sumner, now chairman of the powerful Senate Foreign Relations Committee, did not come. Hence at the end of six months his acquaintances among London editors did not go much beyond Samuel Lucas of the *Star*, Meredith Townsend of the *Spectator*, and Frederick Edge, the American editor of the *Morning-Herald*. In spite of his brother's urgings, he would not degrade himself so far as to become a writer of letters to the London *Times*.

Being the full-fledged London correspondent for pay of a leading New York newspaper did not quite make up for the shortcomings of his position as private secretary. Impatient to make his influence felt, he had to content himself at the beginning with countless menial errands. He was not sure that such might not be his "only duty always," and he gloomily foresaw that his own share "in matters in general" would be "very small." This fit of depression was the precursor of many such moments of disillusionment; fortunately however the pendulum of his moods swung as often toward elation, though never with equal force. As time went by he entered more and more into his father's confidence, much to the chagrin of the official assistant secretary of the Legation, Benjamin Moran, who, jealous of his prerogatives, at last took bitter refuge in his diary: "The two sit upstairs there exchanging views on all subjects and as each considers the other very wise, and both think all they do is right they manage to think themselves Solomons and to do some very stupid things." [4]

Though technically "outside of the Legation," Henry managed at first to get on fairly well with the official staff. Charles Lush Wilson, the former editor of the Chicago *Daily Journal* and a hearty Westerner, was the new Secretary of the Legation. The careerist Moran, whom Charles Francis Adams inherited from his predecessor Dallas, remained eclipsed as assistant secretary. From time to time the ill-matched threesome went about to official functions at which Henry always demeaned himself in a "pleasant and gentlemanly manner," according to Moran, even though they were sometimes driven to amuse themselves "like little Jack Horner in a corner philosophizing over the company." It was a shock to Henry, who had moved with familiar ease in the most exclusive Boston circles, to encounter the exquisite cruelties of British exclusiveness. Keenly aware of his social isolation he regaled his companion with cynical witticisms. "As Henry Adams says," reads one of the appreciative entries in Moran's diary, "after you have bowed to the hostess, made some original remarks to her about the weather, and looked at the family pictures, the stock of amusements is exhausted; unless you find some barbarian present with refinement enough, or, if you please, sufficient confidence in you to present you to a young lady, who will talk, it is a waste of time to remain."

One aspect of British life which simultaneously repelled and yet fascinated him was the elaborate mummery of royal levees. He knew that he ought to despise the invidious distinctions and feudal protocol of court society. It seemed to him a society without a social existence; yet he could not help wishing to shine in it. The official court dress which his father had promptly restored became him "exceedingly well" even in Moran's eyes. At his first presentation to Queen Victoria in June of 1861 he made a very dashing appearance in the naval blue coat with its richly embroidered stand-up collar and gold eagle buttons, white kerseymere vest and knee breeches, white silk stockings and a chapeau with gold ornaments. At his side he wore a fine gilt eagle-headed sword. The pageantry of such occasions was hard to resist. He so far succumbed to their ritual that when, a few years later, the long-suffering Moran proved to him that court etiquette did not authorize him to outrank his father's official secretaries he blushed and vowed he would never go to the Court again.

Part of the malaise that clouded his early months in England was no doubt traceable to his chronic dyspepsia. The prolonged sea-sickness of the ocean voyage had left him easily tired and irritable. At his presentation he became ill with excitement and was frightened "nearly to death" with fear of a "relapse." Equally trying to his temper was the corroding knowledge that almost alone of his friends he had no part in the military drama of the war, this in spite of the fact that only a few months before he had himself expressed the wish to join a "Cromwell-type" regiment, a regiment distinguished, as Macaulay said, by the "austere morality and the fear of God which pervaded all ranks." Now each mail brought news of more enlistments and more exalted military commissions. More than one-third of his college classmates were marching off to glory and military rank.[5] Those companions of his European *wanderjahr*, Nicholas Anderson and Benjamin Crowninshield, already held enviable commissions. For one who was developing a passion for excitement, the disciplined inactivity required of diplomats was almost more than he could bear.

The England to which the young unofficial diplomat came in the spring of 1861 was an England bafflingly changed from the country of his preconceptions. Father and son had expected that British foreign policy would reflect the strong antislavery principles of the ordinary Englishman, who had been bred in the school of William Wilberforce. Neither one was prepared for a foreign policy in which economic considerations played so important a part. If Henry Adams had not been so preoccupied with the day to day incidents of the secession crisis in Washington, he might have detected signs that economic determinism had already begun to pervert the moral professions not only of the ruling classes of the British Empire but also of the American imperialists gathered about Secretary Seward. The realistic trend of British policy had for some time reflected the growing uneasiness of the English envoy, Lord Lyons, in his dealings with Seward; but if Seward perceived this suspicion he did not forewarn Henry's father before their departure. Nor did any inkling of a significant change in English policy filter down to the avid notice of the young private secretary.

The leading English journals had at first sympathetically followed the efforts of the antislavery groups to wrest control of the national government from the South. They rejoiced in the election of Abraham Lincoln and the triumph of the Republican party, seeing in those events the ultimate extinction of Negro slavery. Secession might lead, of course, to armed conflict, but it was universally hoped that the sections would somehow muddle through to a peaceful solution. But as the prospect of armed collision became more and more imminent, idealistic considerations gave way to a growing concern with the probable effect of such a conflict upon British economic interests. Cotton manufacturers seemed fully persuaded that the cutting off of the American supply would have a catastrophic effect upon England. The London *Times* as bellwether of the press began to urge that the North should accept secession as an accomplished fact. It justified its apparent retreat from principle with the argument that the incipient civil war was not really a conflict of right and wrong but a mere contest for political domination. Lincoln's inaugural furnished ammunition for this view, for it appeared to guarantee the indefinite retention of slavery in some form.

For those Englishmen who placed their country's business interest first the developments in Washington were not reassuring. Seward's opinion that the Montgomery government was no government and that secession was a mere rebellion—the view held also by Lincoln—alarmed the realistic British representative in Washington. He carefully transmitted to the Home Government all of Seward's threats about establishing a paper blockade against British commerce and his even more menacing talk of a "foreign war panacea." To the English the view that the war was a mere insurrection would have required all maritime powers to become *de facto* allies of the North in hunting down the Southern privateers. Lord Russell, the Foreign Secretary, spoke his countrymen's passionate desire to avoid such involvement when he cried out in Parliament on May 2, "For God's sake, let us if possible keep out of it." To forestall any compromising acts the Ministry had promptly drafted the Proclamation of Neutrality and issued it on May 13, the day the Adamses landed at Liverpool. It was an eminently safe move and superbly timed. Had Minister Adams arrived before the issuance of the proclamation he might have been obliged by Seward to ask for his passports on the ground that the act was hostile to the United States. Now however that the Proclamation was a *fait accompli* Adams could do no more than object that it was precipitately issued and urge its recall.

In his effort to analyze the ebb and flow of British opinion, official and public, Henry Adams labored under a serious handicap. He did not have as much knowledge of Secretary Seward's intentions as Lord Russell had. Unknown to young Adams and his father, there had gone on in the American Cabinet a struggle for supremacy, one of the most bizarre aspects of which concerned Seward's extraordinary memorandum of April 1, "Some Thoughts for the President's Consideration." The paper outlined an impudent policy toward the Atlantic Powers that amounted to an incitement to war. Lincoln quietly shelved this plan; but, Seward still underrating the president's intelligence, prepared a fresh piece of provocation, the notorious Dispatch Number 10 calling upon Queen Victoria's government to adopt the Northern theory of a rebellion. Fortunately Lincoln required that the dispatch be toned down and directed Minister Adams to use the communication solely for his own guidance and not to deliver it to Russell.[6] The insulting contents of the original version of Number 10 somehow "leaked out" and the terms of the affront were allowed to become public property in England.[7] Little wonder, then, that Henry and his father faced a very general distrust of Seward in England. Even in its emasculated form the dispatch bewildered the American Minister. Henry with youthful forthrightness called the proposed policy "shallow madness" and declared himself "shocked and horrified by supposing Seward, a man I've admired and respected beyond most men, guilty of what seems to me so wicked and criminal a course as this." Not sharing Seward's temporary aberration, the American minister reduced the heated phrases of his chief to a chill innocuousness. He had saved his mission, but for how long?

War of Nerves

Henry Adams took his job of foreign correspondent with appropriate seriousness, regularly sitting down each Saturday in the privacy of his two-room suite on the top floor of the Legation residence to write up his notes on the week's happenings. Before coming to a precipitate close in January, 1862, the series of dispatches continued steadily for nearly eight months, only one being omitted, that for August 17. Unsigned, they appeared under such captions as "Important from England," "American Topics in England," "American Questions in England," or "Matters at London" and were commonly credited to "Our Own Correspondent." [1]

The need for complete anonymity was much more imperative than in the case of the *Advertiser* series, quite apart from the prohibition of the State Department. In the earlier series discovery would at worst have compromised his father's leadership of the Massachusetts delegation. Exposure now might discredit the entire negotiation with the British Ministry and make his father's mission untenable. Under no circumstances did he dare reveal that he had direct access to diplomatic secrets. In an early report, for instance, he declared that the Minister had "had at least one, possibly more than one, interview with Lord John Russell," although as private secretary he could have sworn to the exact number. Nor did he dare tip off the American press to the contents of Dispatch Number 10, though he was frantically alarmed. When the British Government dispatched military reinforcements to Canada, in fear of a Northern coup, he guilefully referred to "rumors current that the President and his Cabinet wanted to bring about a foreign war." The hint was calculated to put the people on guard against the activities of such warmongers as the New York *Herald* "or more reliable authorities." Lincoln and Seward needed to be warned, even if Adams could do so only indirectly, that any proposal for dangerous ventures in power politics would not be kept from the public.

By late September 1861, he feared that he had been telling so many secrets that his "position began to be too hot." To create a "little wrong scent" he dated one report from Leamington and the following one from Glasgow and filled them

with guide book allusions to historical landmarks. His fears of
discovery seem morbidly exaggerated. What hints he threw out
were so cloaked in figurative language that their meaning was
obscure or actually misleading. For example, at that moment
the most delicate pending negotiation related to the offer of
the American Government to adhere to the Declaration of Paris.
An outgrowth of the Crimean War, the Declaration of Paris
abolished privateering and established the right of neutral
countries to trade with belligerents. To "Our Own Correspondent" of the *Times* the Ministry was behaving in a very equivocal
manner toward the American offer. "The English Government
stutters and stammers and trips itself up in the awkwardest
manner in its effort to keep the hems of its garments out of the
mire in which we are stuck, and in consequence look with the
most laughable shyness at the trap which they suspected under
this amicable proposal." Actually Russell and his confreres had
ample reason for suspecting a trap. Seward did intend to maneuver England and France into withdrawing their recognition
of the South as a belligerent. Completely ignorant of this
scheme Henry confided "some of the horrors of the prison
house" to Charles under the most solemn injunction of secrecy.
His revelation amounted to no more than another hint of what
was public knowledge. "This matter has dragged its slow length
along through strange delays, misunderstandings, and discussions that in so simple a matter were very curious and inexplicable."

The position which Henry Adams took on Anglo-American
affairs continued without break the conciliatory policy of the
Times correspondent who preceded him. It assumed that the
English people were in sympathy with the North. The fact
granted, it followed that Americans should do nothing to make
for ill will, no matter what provocation might be offered by a
few pro-Southern English journals. Editor Raymond relied on
him almost exclusively for commentary on the British view of

American affairs. Like his predecessor Adams placed the main emphasis of his news and comment on political and diplomatic affairs. Discursive and gossipy, his columns ranged at will through the sensations and small beer of the week, once Anglo-American affairs were disposed of.

His belief that English public opinion strongly favored the North soon fell victim to the facts. The British public was apathetic. "So far as has been shown as yet," ran his published indictment, "the English people are looking on at the great struggle in which their natural allies are grappling with a powerful enemy, and modern civilization and medieval barbarism stand with their hands on each other's throats, and at the very heat of the contest, they, the bulwark of Liberalism in Europe, have no word of encouragement or hope for the one or the other, and make it their pride to stand neutral. Neutrality in a struggle like this is a disgrace to their great name." Some weeks later, however, when he read the unexpectedly violent American press attacks on the neutral policy of Great Britain, he tried to lay the tempest which he had helped to raise. "The sympathies of the English people are actively with us because they feel, in their cold and practical way, that their true interests lie with the North, and their common sense tells them that the cause of free institutions, their own cause as much as ours, is bound up in the result of our contest."

Nevertheless, as time went on, he reported that "steadily and surely popular opinion is forming here against us" and concluded at last that "sympathy in England or elsewhere is to be won by the sword alone." So far it was the South that grasped that weapon by the hilt, as Ball's Bluff and Bull Run proved. Disregarding an earlier less gloomy estimate, he asserted in November that "within the last six months popular opinion has run steadily against us here." Thereafter the ill will of the English seemed an established fact. What particularly angered him was that the Americans had brought this hostility upon themselves. "How do you suppose we can overcome the effects of the New York Press?" he complained to Charles. "How do you suppose we can conciliate men whom our tariff is ruining? How do you suppose we can shut people's eyes to the incompetence of Lincoln or the disgusting behaviour of many of our volunteers and officers?" Especially exasperating were the meddlesome activities of such "noisy jackasses" as Cassius Clay and Anson Burlingame, who came abroad to help guide public opinion. Publicly more discreet, he promised his readers "to speak in some future letter" of the reasons for the waywardness of British opinion but he did not return to the subject again.

Among the Press his special aversions were the London *Times* and the New York *Herald*. They were most to blame for engendering bad feeling. Almost every dispatch contained an attack on the low motives of the London *Times*. After Bull Run he wrathfully observed: "The *Times* at once came out in a tone so needlessly insulting and so wantonly malignant that no one could doubt any longer . . . on which side its sympathies lie. Now he solaced himself that "the *Times* is not England, nor even the best part of it"; again, that "the *Times* touches nothing which it does not disfigure and states nothing which it does not misrepresent." At last, losing all patience with Editor Delane, he assured his readers that the blackguard London *Times* did not even express the policy of the Ministry and should therefore be disregarded. As for the New York *Herald*, he tried to offset its subversive influence by charging that it was as false a publication as the London *Times*. When the *Herald* published a somewhat garbled account of his father's confidential report to Seward of an interview with Lord Russell, Henry denounced the "pretended report" as the work of "the lowest print that has ever disgraced a great nation." On the other hand, he gratefully acknowledged that the editorials of the London *News* were "sound and determined on the right side." But when the adverse current swelled to a flood, only the unimpressive

Star, the organ of the working class leader John Bright, was left to side with the North.

On occasion he recognized that no simple generalization could fairly describe the tangle of loyalties and interests in England. He was able to distinguish such groups as the Northern sympathizers, the commercial classes, the antislavery and constitutional liberals, the Cotton Interest, the Liberals and Conservatives, the Radical Party, the War Party, the middle and lower classes, and even the politically weak laboring class, but in the heat of righteous indignation he obliterated these distinctions. "I have asserted many times," he wrote, "that the English people dislike us, fear us, wish to see our nation crippled, and our free institutions overthrown." When the Mason and Slidell story broke in London, the rabid hostility of the English press drove all thought of conciliation out of his head and he lashed out with the invective he formerly reserved for Southerners. "The phlegmatic and dogmatic Englishman has been dragged into a state of literal madness, and though not actually riotous, he has lost all his power of self-control. He is seldom well-informed on any but English subjects looked at from a national point of view; he is often sullen, dogged, and unsocial. But in these December days it is worse than all this; it is sheer, downright national insanity, cropping out in the characteristic forms in which his greatness and weakness always proclaims itself." Perhaps this savage criticism was intended to have the therapeutic value of a dash of cold water upon possible English readers; it could not have had much value in sustaining his father's efforts to reconcile the parties.

Apart from several such patriotic outbreaks, he did use his column fairly consistently to discourage attacks upon the English Ministry and upon business and social interests that might be provoked to more active hostility toward the United States. He warned his readers of the danger of embarrassing the present Liberal Ministry, reminding them that it was the Liberal

Party whose ideas "have been the means of protecting the advance of liberal opinions in Europe." He defended Lord Russell as a friend of America and democracy, and extenuated Russell's issuance of the offending Proclamation as due to the pressure of his colleagues in the Cabinet. "So long as Lord Russell is at the head of Foreign Affairs," he affirmed, "I believe America may feel confident that no encouragement will be given to the Slave Power." Palmerston, on the other hand, he portrayed as the evil genius of the Ministry, an "old school Machiavellian" and everyone's "friend," a statesman without moral scruple whose system required of him simply that he "act for the day and let the morrow look out for itself."

For the most part he vigorously skirmished along the propaganda front here attacking the causes of grievances, there playing down sources of dissension. Of first importance was the "amelioration" of the Morrill Tariff. "If we have a great man to direct our country," he hopefully hinted, "he will neglect no opportunity of binding to our side the interests even more than the sympathies of every foreign nation for the present." When the Confederate raider "Nashville" took refuge in an English port after burning the "Harvey Birch," Adams pleaded with his countrymen "not to explode until it appears that we are wronged," for, as he attempted to reassure them, England "will not refuse justice now, when it is clearly on our side." However, the plea was obsolete. Captain Wilkes of the U. S. S. "Jacinto" had already boarded the Royal Mail Packet "Trent" and removed the Confederate envoys, Mason and Slidell. News of the affront to the British flag did not reach London until November 27. Adams's embarrassment was heightened by the fact that in an earlier dispatch he had denounced the rumors of a plan to intercept Mason and Slidell as a canard manufactured by Southerners. "One has to be perpetually on one's guard against the miserable, intriguing spirit of these fellows, who are at the bottom of every contemptible plot." Perhaps because he relished the unconscious irony, editor Raymond allowed the libel to stand.

The "Trent" affair unnerved him. He admonished his audience that the seizure was "a blunder if not a crime." Desperately, he proposed playing "our last and highest card," the immediate freeing and arming of the slaves in order to bring the war to a quick and successful conclusion and thus take the moral initiative away from the war party in England. Seward frowned upon the suggestion as ill-timed and the London correspondent scrambled back to safety. "I am an abolitionist," he reminded his brother, "and so, I think, are you, and so, I think, is Mr. Seward; but if he says the time has not yet come . . . then I say, let us wait. It will come. Let us have order and discipline and firm ranks among the soldiers of the Massachusetts school." Charles, much more conservative on the slavery question than Henry or their father, declined the label and placed in evidence his three recently published *Independent* articles which further developed the thesis that slavery could best be abolished by economic pressure rather than emancipation.

The retreating roar of the "Trent" affair filled Henry's last five dispatches. He pleaded that the British demand for the surrender of the Confederate commissioners should be complied with and complied with "cordially," as was finally done. The right or wrong of the matter was unimportant in the then state of English public opinion; anyway, he comforted his fellow Americans, "We have all eternity to settle our account with England."

Raymond's editorials on Anglo-American affairs generally followed the line set by his London correspondent. In the "Trent" case, however, sharing the popular satisfaction at what was imagined to be England's discomfiture he altered one of the dispatches, much to young Adams's annoyance. "There's the New York *Times*," he grumbled to Charles, "which I warned only in my last letter against such an act and its consequences; and now

I find the passage erased, and editorial assurances that war was *impossible* on such grounds. Egad, who knew best, Raymond or I?" He had little additional cause to complain, for Raymond tampered with very few other pieces. The early article scornfully rejecting the rumors of American designs on Canada was cut to a mere quarter column. The November 22 column reporting British reactions to the correspondence between Lord Lyons and Secretary Seward survived as a confused medley of scraps. Beyond those instances, Henry Adams was allowed to speak his mind.

Extracts from a Private Diary

In Boston, Charles was doing his part to direct public opinion through the columns of the *Transcript,* the *Courier,* the *Advertiser,* and the New York *Independent.* The letters that sped back and forth across the Atlantic as swiftly as the new Cunarders could carry them were filled with debate on the conduct of the war, the wisdom of Seward's diplomacy, and the state of opinion in America and England. Once, carried away by his own rhetoric in defense of Seward—the foreign war delusion having passed—Henry concluded "If you think the above worth printing, send it to Charles Hale." The letter went unprinted, and Henry was left to reflect dismally on his brother's superiority as a writer. The *Courier* articles seemed so "devilish good," in spite of the fact that they inclined heretically toward Sumner's "ideas of Washington affairs" rather than Seward's, that Henry felt "blue for a day," thinking of his own "weak endeavors."

After the exasperating defeat of Bull Run, he demanded that his brother obtain a commission for him. "I cannot stay here now to stand the taunts of everyone, without being able to say a word in defence." This craven plea was too much for Charles. "Go to work at once in England with all your energy and force your way into magazines and periodicals there and in America," he exclaimed. "Look into the cotton supply question . . . and try to persuade the English that our blockade is their interest. . . . Then write to the *Atlantic* of the way fighting America appears in English eyes, of her boasting and bragging, her running and terror . . . Here is your field, right before your nose. . . . Don't talk of your connection with the legation to me; cut yourself off if necessary from it and live in London as the avowed *Times* correspondent and force your way into notice of the London press that way . . . Free from the legation you could earn a living by your pen in London and be independent, busy, happy and eminently useful."

Nor did Charles let the matter rest there. In his next letter he offered to turn over his notes for an "elaborate article on this cotton supply question" and gave copious suggestions on how to work them up. With equal vigor, he probed at a central weakness in Henry's literary style. "You always affect in writing too much calmness and quaint philosophy. That will come to you in time, but you do it now at the price of that fresh enthusiasm which is the charm of young writers. If you write now, write as if you were pleading a cause and too much interested to be affected."

As might have been expected, Henry resisted being pushed into a new project with such velocity. It seemed to him that such an article would be superfluous because it was already axiomatic in England that continued dependence on the American cotton monopoly was a curse. Moreover, he felt confident that the Ministry would not be induced to touch the blockade question. It was "a generally acknowledged truth" that by spring the "cursed monopoly will be broken and with it the whole power of the south." In any case he had his own projects "in another direction." Nothing came of these "bubbles" and three weeks later he was at work on still another lead: "My great gun is the Manchester one. Tomorrow evening I start with a pocketful of letters for Manchester to investigate that good place . . . My present plan is to report with as much accuracy as possible all my conversations and all my observations, and to send them to you. Perhaps it might make a magazine article; except that it should be printed as soon as possible. If I find that I can make it effective in that form, I shall write it out and send it to you for the *Atlantic*. If not, I shall contract it and send it to you for the *Advertiser* or *Courier*." As the report arrived "just too late" for the January *Atlantic* Charles reluctantly "carried it to the *Courier*" to pay off a debt to that "low-toned and semi-treasonable sheet." Published on December 16 under the heading "A Visit to Manchester—Extracts from a Private Diary" the prudently unsigned two and a half columns of type reported five days of interviewing among Manchester industrialists, supported by a hasty study of the comparative value and availability of non-American cotton.[1]

A chief purpose of his errand was to ascertain the "feeling of the solid people of Manchester towards the North." There could, of course, be no question of the friendship of "the radical party, the Brights, and the Cobdens, who have large influence." Pooling all the opinions he heard, he concluded that "we need fear no active hostility from Manchester." He discovered that both he and his brother had been mistaken about the reason for the shutdown of factories. Southern propaganda had successfully persuaded the public that the shortage of raw cotton was beginning to be felt.[2] Actually, as was explained to Adams, the shortage could not begin to exert its influence until the following year. The spinners had so thoroughly anticipated the war that the market was now glutted with finished goods and prices were so demoralized that some mills had found continued operation no longer profitable. In one mill "about a quarter of the spindles were silent, and, as they told me, a corresponding number of the operatives discharged, to starve as best they might." He noticed that the operatives still at work "were very dirty, very coarsely dressed, and very stupid in look; altogether much inferior to the American standard."

The Cotton Supply Association was aggressively seeking substitutes for the American long staple, but without success so far. The mills had already experimented with India cotton and at least one manufacturer thought the fabrics could replace "all but the finest Americans, at ordinary prices." In examining the stuffs themselves Henry could detect no visible differences. The experience of India convinced him that the South did not have a "natural monopoly" on a cotton climate. If the blockade held, "spring will find England nearly independent of America for this article, and we shall see the steady advances of a great revolution in the world's condition." The implication was obvious that the North must redouble its efforts to maintain the blockade and hurry the war to a conclusion before the English manufacturers should feel the pinch.

One of his incidental comments had a fateful influence upon his career as a journalist. Perhaps to enliven the dreary technical aspects of his report, he interjected a comparison of London and Manchester society.

Manchester society seems to me much more like what one finds in American cities than like that of London. In Manchester as in America it seems to have fallen, or be falling, wholly into the hands of the young, unmarried people. In London the Court gives it dignity and tone, and the houses into which an admission is thought of most value, are generally apt to slight dancing. In Manchester, I am told, it is still the fashion for the hosts to see that their guests enjoy themselves. In London the guests shift for themselves, and a stranger had better depart at once so soon as he has looked at the family pictures. In Manchester one is usually allowed a dressing room at an evening party. In London a gentleman has to take his chances of going into the ball room with his hair on end or his cravat un-

tied. In Manchester it is still the fashion to finish balls with showy suppers, which form the great test of the evening. In London one is regaled with thimblefuls of ice cream and hard seed cakes.

The anonymous attack upon his London hosts relieved the long pent accumulation of his social grievances; but it proved an unlucky revenge. The editor of the *Courier,* pleased to receive a new contribution from the young correspondent whom he had so fulsomely praised a year and a half earlier, disregarded his instructions and blurted out the secret of its authorship.

The interesting Diary at Manchester, on the outside of today's Courier, *we feel at liberty to say, is written by Mr. Henry Adams, the son of our Minister to Great Britain. This accomplished young gentleman has been for some time abroad. This Diary shows that he has by no means degenerated from the hereditary ability of his family,—which now for four generations has either fulfilled high expectations, or, as in the present case, has given promise of future distinguished usefulness to the country.*

Here indeed was a morsel for the London journalists. To Henry's "immense astonishment and dismay" he found himself "sarsed through a whole column of the *Times*" [3] and made the laughing stock of England. The editorial described him as a "Special Commissioner" who, having been "despatched on a voyage of discovery . . . proceeded to take soundings on this unexplored and dangerous coast." The writer blandly denied, of course, that Adams's report correctly reflected Manchester opinion and then turned patronizingly to the comment on London society. What Adams needed was a less limited social experience. "Let him but persevere in frequenting *soirees* and admiring 'family pictures' . . . and we shall not despair of reading some day, a new diary in the Boston *Daily Courier,* wherein the *amende honorable* will be made to the gay world of our metropolis." Young Adams might have faced out a downright hostile criticism, but such condescension made him writhe in agony.

The exposé filled him with a panicky fear that his authorship of the articles in the New York *Times* might also be uncovered. He must stop that series at once. But how? A letter from the Legation might tip off one of Raymond's underlings. Only Charles could help him. He asked his brother to write to Raymond, "without mentioning names," explaining "why his London correspondent has stopped for a time." At all costs their father must not learn of his connection with the New York *Times.* "The Chief," wrote Henry, "bears this vexation very good-naturedly, but another would be my ruin for a long time."

The London press was not yet done with him. On the day following the *Times* leader, the *Examiner* impaled him on its wit, gibing at the "frightful risk" he took "of going into a house with his hair on end and his cravat untied." As for his objections to the "hard-seed-cakes and thimblefuls of ice-cream, . . . this should be a caution to all persons giving parties . . . to be more careful about their cakes, the softness, and the seeds thereof . . . That hard seed-cake runs through and embitters all the young gentleman's reports of us." The critic seized upon Adams's assertion that "the slave power may again be curbed to its due position in politics" to argue that it showed Adams did not really desire nor could indeed imagine "the extinction of slavery."

Young Adams was too crestfallen to think of defending his position. His sole thought was to make himself inconspicuous. A bolder man might have cut himself off from the Legation then and there, as Charles had once advised, and have forced the discussion of the real issues. But at heart Henry was

no brawling controversialist and the impulse to withdraw after
the first assault became a fixed habit. The episode was not forgotten as completely as he afterward liked to think. Benjamin
Moran's Diary records a sustained pleasure in the recollection.
Much later, for example, when Adams showed great resentment
at not having been invited to the Guildhall ball, Moran
scrawled in his diary, "This outrage should be avenged by the
nation . . . To cut him is almost, though not quite, as bad
manners as making a joke of your friends' hospitality for the
public press."

The public attacks so unnerved him that he foreswore further journalism. He suspected that Charles would not approve
his retreat in the face of danger. To forestall renewed criticism
he explained: "I have wholly changed my system, and having
given up all direct communication with the public, am engaged
in stretching my private correspondence as far as possible."

His new method of operation is illustrated by a letter addressed to Thurlow Weed, then in London as one of Seward's
emissaries. Apparently the diplomatic correspondence relating
to the "Trent" affair was about to be published. Delay had in-
tensified the painful suspense. War had never seemed closer.

Friday Night

My dear Sir

*Will you do me the favor to read the enclosed? It seems to
me to be advisable to give, if possible, an immediate direction
to public opinion on the appearance of the documents— This
letter is intended as a keynote, but of course I would not offer
it for publication without your approval— Nor am I at all disposed to press it— If you object to it* in toto *or in* parte *I would
not on any account have it printed—*

*If printed, however, it must of course be done without the
knowledge of my father, as that would commit him to it. As I*

*do not wish, myself, to be known now as a writer for the press,
I would prefer to have its origin kept secret.*

   Excuse my officiousness, I pray; and believe me
                                    Very truly Yrs
                                    Henry Brooks Adams

The enclosure, like so many of its predecessors, sank impotently
into the political void. As Henry said, he tried to be useful to
Weed "but without much result." [4]

The break in what had become a well-established literary
routine touched him more deeply than he at first would acknowledge. Self-mistrust, never far below the surface of his
mind, reasserted its claims, and his misgivings overflowed into
his letters to Charles. On the deck of the transport lying off
Beaufort, South Carolina, the young lieutenant greeted Henry's
vaporings with measureless scorn. "Fortune has done nothing
but favor you and yet you are 'tired of this life.' You are beaten
back everywhere before you are twenty-four, and finally writing philosophical letters you grumble at the strange madness of
the times and haven't even faith in God and the spirit of your
age. What do you mean by thinking, much less writing such
stuff? 'No longer any chance left of settled lives and Christian
careers!' "

Completely routed, Henry poured ashes upon his head. "I've
disappointed myself, and experience the curious sensation of
discovering myself to be a humbug. How is this possible? Do
you understand how, without a double personality I can feel
that *I* am a failure? . . . You are so fortunate as to be able to
forget self-contemplation in action, I suppose: but with me, my
most efficient channels of action are now cut off."

Vague plans flitted through his fancy, but his heart was not
in them. He was rediscovering the "morbid self-reflections
which always come from isolation in society." [5] On the eve of

the first anniversary of the war, he wrote *de profundis:* "I feel ashamed and humiliated at leading this miserable life here, and since having been blown up by my own petard in my first effort to do good, I haven't even the hope of being any more use here than I should be in the army." No amount of wishing could produce the god in the machine. There was nothing for it but to make a new and more philosophical approach to the choice of a career.

Chapter Five

A Golden Time
1861–1868

Young England, Young Europe

ONCE when revisiting London in the early Nineties Henry Adams "lapsed from his usual cynical manner" to give advice to a young companion, Lloyd C. Griscom, who was just beginning a career as a diplomat—advice which Griscom never forgot. Looking back on his own initiation many years before, Adams said, "It was a golden time for me and altered my whole life . . . You're in a remarkable position now. You've every opportunity to make friends that will influence your entire career. Be sure to keep your head and get the most you can out of it." [1] In that moment of musing insight, he evaluated his English experience much more sympathetically and fairly than when he again reviewed it in *The Education* a dozen years later. His later judgment was cut from the same mortuary gauze with which he draped his memory of Harvard: "He was in dead-water and the parti-colored fantastic cranks swam about his boat, as though he were the ancient mariner, and they saurians of the prime . . . He knew no more in 1868 than in 1858 . . . He could see only one great change, and this was wholly in years."

When he reached England in 1861 he was twenty-three years old; he was thirty when he sailed for home in 1868. That interval marked the transformation of the provincial young man who had once been homesick for Quincy and Boston to the cosmopolite "with an English cut to his jib" [2] who gravitated to a capital "by a primary law of nature." He brought ashore with him from the "Niagara" a baggage of ideas that bore the labels of Quincy, Boston, and Harvard College, ideas that in a larger world were often indistinguishable from prejudices. His two years on the Continent had scarcely touched these; nor had they shaken his faith in a priori moral and political principles. In the society of Congressmen and lobbyists, of politicians, lawyers, and editors, the society in which he had moved in Washington, the impending war had displaced practically all other interests. As London correspondent his view had widened to include international politics and diplomacy, but it continued to be preoccupied with the parochial concerns of a newspaperman. In a sense the abrupt change worked by his exposure also freed him from those narrow concerns, for as his work as a propagandist became attenuated he no longer had so strong a motive to identify himself exclusively with his father's diplomatic and political interests. New influences could now enter his mind more freely and make their challenge felt—new trains of thought, new lines of literary interest.

His English associations were prescribed, at least at the beginning, by an almost iron law of affinity. The highest aristocracy maintained a stony reserve toward the representatives of the Union cause and did little to conceal their preference for the Confederacy. Since his social standing derived from his position as son and private secretary to the Minister, Henry Adams fell equally under the ban. Though themselves members of an American aristocracy, the Adamses were driven into the arms of the largely unaristocratic English liberals and Radical reformers. The names of such enlightened sympathizers as William E. Forster, John Bright, Nassau Senior, Monckton Milnes, Thomas Hughes, Richard Cobden, and Robert Browning appear early in Adams's London Date Book. These men were part of a circle which included figures like Sir Charles Lyell, Goldwin Smith, John Stuart Mill, Leslie Stephen,

Thomas Huxley, Frederic Harrison, Francis Newman, and Harriet Martineau, all of whom gave aid and comfort to the American Legation. Young Adams struck up a friendly acquaintance with most of the influential friends of the Union and basked in the sunshine of their sympathy for his country. But living in the shadow of his father's distinction as his father and his father's father had done before him he could not yet make his mark in spite of his keen wit and manifest good breeding.[3]

Except for John Bright's mention of a conference which included Thurlow Weed and Henry Adams,[4] none of the many published memoirs and biographies of his English acquaintances during this period make any allusion to him. It was his fortune—or misfortune—to be thrown into the company of eminent persons generally older than he and he had to endure the constraint imposed by age upon youth. On one occasion, however, he was bold enough to debate the subject of free trade with John Stuart Mill "to whom he took particular pains to be introduced." He thoroughly appreciated his opportunities even though he was usually cast in the role of listener. At one "royal evening" at which the company was of "the earth's choicest," Robert Browning and Bulwer-Lytton fell into a discussion on fame. "It was curious," Adams reflected, "to see two men, who of all other, write for fame, or have done so, ridicule the idea of its real value to them." Another time he was a guest at "a pleasant dinner" at which Charles Dickens, John Forster, the much-admired biographer of Oliver Goldsmith, Louis Blanc, the eminent historian and politician-in-exile, and "other distinguished individuals were present." One of the most memorable experiences was the intimate gathering at Fryston Hall which brought into his ken one night young Algernon Swinburne, whose genius blazed upon him like a comet and was gone.

He was soon quite ready to concede that "society in London certainly has its pleasures," as for instance when he dined at the home of the Duke of Argyll, a Liberal leader in the House of Lords, and there met Charles P. Villiers, a member of Parliament and a prominent Free Trader, Charles Brown-Sequard, the celebrated neuropathologist, Professor Richard Owen, the paleontologist, and Lord Frederick Cavendish, private secretary to the president of the privy council. Again after returning from a breakfast with William Evarts, Cyrus Field, Cobden, and others, he wrote in a lofty vein: "So we go on, you see, and how much of this sort of thing could one do at Boston!" He continued to insist, however, that these exquisite moments were islands of delight in the "vast nuisance and evil" of fashionable society, and gave only partial assent to Lothrop Motley's dictum that this was "the perfection of human society." [5] Perhaps a tiny fraction of the evil was contributed by the sharpness of his own tongue. Once at a party, Moran tells us, "Lady Holland remarked to Henry Adams that we Americans are unmistakable, and pointing to us, said 'See those gentlemen—anybody would know the American from the Englishman.' 'Which of them, Lady Holland, is the Briton?' said Adams. Her ladyship seemed somewhat piqued, and said, 'Why the dark-haired one of course.' 'It happens,' remarked he, 'that that is Mr. Moran, the Secretary of the Legation, a native Pennsylvanian, and your Yankee is every inch an Englishman, and has not been in the United States.' Lady Holland dropped the conversation."

Life in a great world capital was indeed different from the provincial amenities of Boston. The hundreds of entries in his appointment book, which he kept with widely varying zeal from 1861 to 1868, tell of an almost endless succession of interviews, calls, breakfasts, dinner engagements, balls, teas, and at homes; and the record glitters with the names of Englishmen and their wives who were helping to make him a man of the great world, anglicizing his speech and manner almost in spite of himself. Among the more frequently recurring entries are the names of Lady Palmerston, Lady Russell, Sir Charles Lyell, Sir

Emerson Tennent, Sir Henry Holland, Lady Holland, Richard Monckton Milnes (Lord Houghton), William E. Forster, John Bright, Nassau W. Senior, Thomas Baring, the Duchess of Northumberland, the Countess of Harrington, the Countess of Derby, the Duke of Devonshire, Lady Stanley of Alderly, Lady Waldegrave, the Duchess of Somerset, Lady Rich, Lady Lampson, Lady Goldsmid and Sir Francis Doyle. Men like Charles Milnes Gaskell, Sir Robert Cunliffe, Russell Sturgis, and Francis T. Palgrave figured as intimates. It appears that Robert Browning, solitary after the death of his wife in 1861, occasionally came to dinner. A brief entry indicates that when Dr. Palfrey visited London in October of 1867, the historian James A. Froude was invited to meet him at the Adamses.

The Legation attracted not only Englishmen of importance but also all visiting Americans great and small. Many of his countrymen whom Adams now met or with whom he renewed acquaintance were destined to play an important part in his life. American diplomats on their way to European posts in 1861 almost invariably stopped at the Legation to confer with Henry's father and Henry was commonly a party to such conferences. As a result he had the chance to meet Carl Schurz, the new Minister to Spain; John Lothrop Motley, Minister to Vienna; James S. Pike, Minister to Holland; Bayard Taylor, Chargé d'affaires at St. Petersburg. Charles Hale, his former employer on the *Advertiser,* likewise paused on his way to Egypt as American consul. William Dean Howells, appointed to the consulate at Venice, came in seeking to be married in the Legation. Secretary Moran, who thought him a "sleek, insipid sort of a fellow," haughtily turned him away before he could meet either Henry or his father. Questions of government finance brought experts like Abram S. Hewitt and Robert J. Walker into the Adams family circle. William M. Evarts and Thurlow Weed came as "roving diplomats" and proved the most acceptable of Seward's horde of special agents. Henry

Adams particularly valued Evarts' instruction and with him he discussed "affairs at home and philosophic statesmanship, the Government and the possibility of effectual reform."

The trickle of American visitors grew to a flood as the Civil War came to a close. Among the neatly recorded names of dinner guests one encounters a number of special significance: Major John Hay, David A. Wells, Congressman James A. Garfield, Congressman James G. Blaine, Colonel Oliver Wendell Holmes, Senator John Sherman, Major General Schofield, Phillips Brooks, Charles Deane, John Gorham Palfrey, and one Dr. Hooper of Boston and his daughter Marian, who was one day to become Adams's wife. Hay has left a glimpse of one of the parties. "We tore our friends to pieces a little while. Motley got one or two slaps that were very unexpected to me. Sumner and his new wife were brushed up a little." [6] There were of course many others who called at the Legation in the "slum," as Secretary Moran disdainfully referred to the premises at 147 Great Portland Street to which the Legation office had been removed, or who enjoyed the pleasanter hospitality of the residence at 5 Upper Portland, notables like Professor Francis Bowen, Richard Henry Dana, Jr., Kate Chase Sprague, Henry J. Raymond, James Gordon Bennett, George Smalley, William Lloyd Garrison, Henry Ward Beecher, and Julia Ward Howe.

Nothing depressed Adams so much, however, as the difficulty of making friends among the English aristocracy. The complaint is a recurring burden of his letters to his brother. It was no triumph to be welcomed by the friends of the North, for they were but a militant minority among the upper classes. And the fact that he was civilly tolerated by the Court entourage only heightened his annoyance. The "perfect exclusiveness" of London society, nevertheless, challenged assault. Henry's parents, who were understandably eager that he should get as much as possible out of his stay abroad, did what they could to smooth the way for him. If tickets were scarce for a state occasion,

Henry managed to attend, sometimes at the expense of his socially ambitious colleague Moran. After one reception Moran
sneered: "He was there pretending that he disliked it and yet
asking to be presented to everybody of note." His social apprenticeship extended beyond England. In December of 1862 he
was sent as a special messenger with a present for the King of
Denmark, and a year later he traveled to Paris to be presented
in full dress to His Imperial Majesty Napoleon III. Whatever
else he might achieve in European society he had at least met
three reigning monarchs face to face, sovereigns who easily
outshone the drab tenants of the White House.

Slow as his social progress seemed to be, the tide finally began to turn after the capture of New Orleans had created
a respect for Northern might that subsequent reverses could not
wholly quench. In March 1863, Monckton Milnes put him up
for the St. James Club, the proposal being "seconded by Lawrence Oliphant, a thorough anti-American," as Adams called
him. In the following month he met Charles Milnes Gaskell,
son of James Milnes Gaskell, Member of Parliament for Wenlock in Yorkshire. Within a short time he was on terms of the
greatest intimacy with the young man and the lively sportive-
ness of his letters to his new friend reflects how deeply grateful
he was for the chance of dropping his diplomatic guard. He
was now discovering that "individually" the English "are very
like ourselves and are very pleasant people." Through Gaskell
he soon met Francis Turner Palgrave, Gaskell's brother-in-law,
and through him, Thomas Woolner, Palgrave's protégé and
other figures of London's intellectual circles. He also became
acquainted with Henry Reeve, the long-time editor of the *Edinburgh Review*. By the end of 1863, alluding to Arthur Clough's
poetry, which had just been edited by Palgrave, he was able
to write: "Young England, young Europe, of which I am by
tastes and education a part, the young world, I believe, in every
live country, are reflected in Clough's poems very clearly."

The New Science

One of the stanchest Northern sympathizers who early allied himself with the Legation was Sir Charles Lyell, the foremost geologist in England. In the dark days that followed the defeat at Ball's Bluff and the blunder in the "Trent" affair, Lyell became a frequent visitor and was soon on terms of intimacy with the entire family. He was then busy on the manuscript of his *Antiquity of Man* and moving toward an acceptance of the theory advanced in his friend Darwin's recent *Origin of Species*. Darwin who "saw more of Lyell than any other man" has described the qualities which must have made Lyell's presence a perpetual stimulus to young Adams. "His delight in science was ardent, and he felt the keenest interest in the future progress of mankind." Through Lyell, young Adams was inevitably drawn into the passionate struggle over Darwinism. Lyell could tell him at first hand of the dramatic scene at the Oxford meeting of the British Association for the Advancement of Science in June of 1860 when, as Leonard Huxley recalls, in the "open clash between Science and the Church," Thomas Huxley routed Bishop Wilberforce after his scoffing attack on Darwin.[1]

The clamor still echoed in the very month that Adams reached England, for Huxley had just concluded his daring lectures on "The Relation of Man to the Rest of the Animal Kingdom" and was in the midst of his violent controversy with Sir Richard Owen, who fiercely opposed the thesis of the *Origin of Species*. The level of public controversy over the book may be gathered from Huxley's picturesque comment in the *Westminster Review:* "Old ladies of both sexes consider it a decidedly dangerous book . . . while every philosophical thinker hails it as a veritable Whitworth gun in the armory of liberalism." As a subscriber of the *Atlantic Monthly* Henry had already read Asa Gray's defense of Darwin against the contemptuous criticisms of Louis Agassiz. Arguing "that Darwin's hypothesis is the natural complement to Lyell's uniformitarian theory in physical geology," Gray concluded that "the rumor that the cautious Lyell himself has adopted the Darwinian hypothesis need not surprise us." Though "not disposed nor prepared to take sides for or against the new hypothesis," he acknowledged that "the imperfection of the geological record" was perhaps the weightiest of the unanswerable objections to Darwin.[2]

To maintain himself in controversy on such ground, Adams would have had to pore over the arguments and counterarguments in the pages of the *Natural History Review*, the *Quarterly Journal of the Geological Society*, and perhaps the *Proceedings* of the Royal Society. At the very least he found himself obliged to read the work that had touched off the great debate, the *Origin of Species* itself, as well as its lesser companion, the *Voyage of the Beagle*. Stimulated by these seminal ideas he now found practical use for Agassiz's instruction and like many other young intellectuals dabbled in gentlemanly geology, hunting for fossils in the Wenlock Edge district of Yorkshire near the estate of his new friend, Charles Milnes Gaskell.

It is hard to exaggerate the stir caused by the scientific discoveries of the mid-century. In 1859, for example, Professor Joseph LeConte addressing the American Association for the Advancement of Science declared that the principle of "the correlation and conservation of force must be looked upon as one of the grandest generalities in modern science." Even earlier, Americans who read the *American Journal of Science and Arts* had seen Rudolph Clausius's epochal article "On the Application of the Mechanical Theory of Heat to the Steam Engine." It was translated from the German by Professor Wolcott Gibbs, who later became a colleague of Adams at Harvard. The 1861 convention of the British Association was described by David A. Wells in the Boston *Annual of Scientific Discovery* as one of its most exciting meetings. The subjects which kindled most discussion were "The Origin and Antiquity of Man" and "Iron-Plated Ships." Thermodynamics also had its share of attention. Thomson in discussing the age of the sun's heat affirmed that mechanical energy obeyed a "universal tendency to its dissipation, which produces gradual augmentation and diffusion of heat, cessation of motion, and exhaustion of potential energy through the material universe."

In spite of the new authority with which science now spoke, the reservations voiced by Adams in his Class Day Oration still survived; but they now took a new direction. In April of 1862 the British Government carried out tests which tended to show, according to Henry, that "their iron navy, and their costly guns" were "all utterly antiquated and useless." The consequences both dazzled and alarmed him. "You may think all this nonsense," he forewarned Charles, "but I tell you these are great times. Man has mounted science and is now run away with. I firmly believe that before many centuries more, science will be the master of man. The engines he will have invented will be beyond his strength to control. Some day science may have the existence of mankind in its power, and the human race commit suicide by blowing up the world." Despite this horrendous prospect, a prospect to be brought infinitely closer in less than a hundred years, he succumbed so completely to Lyell's influence that in February of 1863 he confided to Charles: "I have serious thoughts of quitting my old projects of a career, like you. My promised land of occupation, however, my burial place of ambition and law, is geology and science. I wish I could send you Sir Charles Lyell's new book on the *Antiquity of Man,* but it wouldn't do very well for camp reading." This was the new book in which Lyell finally recanted his belief in primordial creation. It was warmly welcomed by such friends of the Adamses as John Bright who found it "deeply interesting" and Stopford Brooke who was impressed by Lyell's effort "to state the facts without prejudice." [3]

The leaven of the new science gave fresh impetus to the philosophical speculations of the Radical group whose support of the North made the Legation a center of common interest and it was with these brilliant young men that Adams most eagerly sought acquaintance. Their interest in science was, however, much like his own, only incidentally scientific. What stirred their imaginations was that aspect of science that always bewitches the layman, the prodigious and the spectacular. Recent research in spectrum analysis suggested to Stopford Brooke, for example, "the possibility of science becoming the ground and subject of a new school of poetry," a possibility once anticipated by Wordsworth. In such an atmosphere there was little likelihood that Adams would adopt the plodding scientific caution exhibited by the members of the Philosophical Club. His imagination responded with more excitement to the implications of the immense firepower of the new Armstrong naval rifle. And like Brooke, Thomas Woolner, and John Richard Green, he found more solid satisfaction in hunting the tangible Pteraspis in the Silurian horizon of Wenlock Edge. Geology and paleontology were then in a sense front page news and had an immediate theological bearing; and the young men were more interested in arguments and analogies than true scientific inquiry.[4]

Hence, as he mastered "the easy and mechanical work" of his official duties, Adams turned to the study of the writers who were seeking to interpret the new movement in science and philosophy, a work which would be much more "on his own account" with the chances "indefinitely against one's ever succeeding in bringing the results into action." The drift of this study is reflected in the following example of what he termed his "unpractical experimento-philosophico-historico-progressiveness":

The truth is, everything in this universe has its regular waves and tides. Electricity, sound, the wind, and I believe every part of organic nature will be brought some day within this law.

But my philosophy teaches me, and I firmly believe it, that the laws which govern animated beings will be ultimately found to be at bottom the same with those which rule inanimate nature, and as I entertain a profound conviction of the littleness of our kind, and of the curious enormity of creation, I am quite ready to receive with pleasure any basis for a systematic conception of it all. Thus (to explain this rather alarming digression) as sort of experimentalist, I look for regular tides in the affairs of man, and of course, in our own affairs. In every progression, somehow or other, the nations move by the same process which has never been explained but is evident in the ocean and the air. On this theory I should expect at about this time, a turn which would carry us backward.

Having soared thus far into the realm of scientific history for Charles's benefit, he suddenly punctured the thought: "The devil of it is, supposing there comes a time when the Rebs suddenly cave in, how am I to explain that!" In deference to his realistic elder brother he adopted a tone of playful mockery, but he did not feel the same compulsion when he visited Gaskell at Wenlock Abbey. There, he said, "I held forth in grand after-dinner eloquence, all my social, religious and philosophical theories."

A tremendous new organon was coming into being that was to shape the course of Western thought for at least a half century, and Henry Adams was irresistibly drawn into that field of attraction. No realm of knowledge could claim exemption from the scrutiny of the new science. Among the world capitals, the London in which Adams lived constituted a chief center of this renaissance. In the natural sciences Charles Darwin was of course the main figure in the group which included Sir Charles Lyell, Sir Roderick Murchison, Joseph Hooker, Thomas Henry Huxley, John Tyndall, and Hugh Falconer. The revolution in physics stemmed from the researches of such men as Sir William Thomson, Sir Charles Wheatstone, Michael Faraday, James Clerk Maxwell and their Continental colleagues, Rudolph Clausius and Hermann von Helmholtz. Mathematicians like Sir Charles Babbage and George Boole, and chemists like Sir William Crookes and Sir Edward Frankland were opening up additional areas of exploration. Adams could have chosen no better places to pick up the social and political implications of the new science than at Fryston Hall, where Richard Monckton Milnes enjoyed collecting celebrities and controversies, or the stimulating breakfasts of Lord Holland in London. In this intellectual climate he was fully able to satisfy his desire for "a systematic conception of it all."

The triumvirate of writers whose works were best calculated to gratify his wish were Auguste Comte, Herbert Spencer, and Henry Buckle, all of whom expected to reconstruct society according to rational and scientific principles. The meetings of the recently founded Social Science Society in London gave promise that such reconstruction would be the work of the new science of society. Frederic Harrison succinctly put the question to himself: "Whether the facts of human nature and society are capable of scientific treatment is the question upon which the course of all future thought must depend." As a Comteist he adopted the hypothesis that they were. When Mill's analytical study *Positive Philosophy of Auguste Comte* was first published in 1865 Adams was one of the first to read it. The impact of this work upon his questing mind may be surmised from the effect it produced upon his brother Charles who was then visiting in England. "That essay of Mill's," said Charles, "revolutionized in a single morning my whole mental attitude." For the benefit of these intellectual collaborators Mill expounded Comte's theory of the three states through which human conceptions pass: the theological, the metaphysical, and the positivist. Social phenomena needed to be reinterpreted according to the scientific principles of the positivist stage. The revolutionary proposal suggested the possibility of an impersonal social physics freed from the tyranny of "all the great social organizations," as Adams later called the vested interests of society. In the light of his subsequent interest in "scientific history," the most arresting passage in the essay reads as follows: "Foresight of phenomena and power over them depend on knowledge of their sequences, and not upon any notion we may have formed respecting their origin or inmost nature." The laws of the new social physics, according to Mill's paraphrase, were to be found in "the constant resemblances which link phenomena together," and "the constant sequences which unite them as antecedent and consequent." Here was assurance that a science of prediction was possible. To the young philosophical statesman with a taste for prophecy the prospect must have been alluring. As he says in *The Education*, "he became a Comteist, within the limits of evolution." [5]

If Comte identified the conditions of such a science, Herbert Spencer first formulated its hypotheses. Spencer, who was one of the philosophers cultivated by Adams's friends at Fryston Hall, sought in his *First Principles* to "interpret the detailed phenomena of Life, and Mind, and Society, in terms of Matter, Motion, and Force." He asserted that "the general law of the transformation and equivalences of forces" holds true not only of physical forces, but of vital, mental, and social forces as well. Human society tends to pass through successive states of equilibrium from a homogeneous into a heterogeneous state with a concomitant loss of motion. Not only society but the solar system itself was dissipating its force as an incident of the universal process of equilibrium. Society must therefore ultimately be extinguished.[6]

If we look ahead no farther than the *History of the United States*, the great achievement of Adams's middle life, the parallels between Spencer's identification of the social and physical processes and Adams's seem more than a coincidence. Adams's early letter to his brother concerning the "regular waves and tides" in the universe impresses one as a good offhand summary of Spencer's theory of periodicity, and the adoption in the *History* of the notion that social energy like an Alpine stream descended through history to be lost finally in the ocean of democracy would seem to add confirmation. One should also note that in the early Seventies in a review of Henry Maine, Adams was to give Spencer a place beside Darwin.[7]

The generalizations of Comte and Spencer were matched in the field of history by Buckle, who attempted in his *History of Civilization in England* "to accomplish for the history of man something equivalent, to what has been effected by other inquirers for the different branches of natural science." Buckle pointed out that history had suffered for want of historians comparable in intellect to a Kepler or a Newton, a comparison that Adams subsequently made his own. He urged that the physical sciences be applied to history because "it is certain that there must be an intimate connexion between human actions and physical laws." The historian must consequently seek to determine the laws of mind and the laws of matter. These daring generalizations, like those of Comte and Spencer, which they resembled, Adams thriftily put aside for future use. They too were to leave their mark upon the conception of the *History*, especially of the great opening chapters. Yet along with his tentative acceptance of these grand generalizations there also lurked the corroding doubts suggested by such men as Francis Bowen and James A. Froude. Bowen in his 1861 review of the second volume attacked Buckle's theory as "gloomy and scandalous," so laden with fallacy as to be "simply absurd." Froude attacked Buckle with such violence in a lecture before the Royal Institution in 1863 that John Morley reproved the obscurantist hostility in an editorial in the *Fortnightly Review*.[8]

Political Acolyte

Adams turned toward "men like Thomas Hughes, and his associates," with a sense of relief, for these men created an island of sympathy in a hostile London. Hughes, like Lyell, was soon made to feel at home at the Legation. His "muscular Christianity" may not have been wholly congenial to Adams's tastes, but he introduced him "to a number of very pleasant acquaintances" at a London literary club to which Monckton Milnes had procured him an invitation. One such may have been Hughes's good friend Frederick Maurice, who with Hughes and Charles Kingsley had inaugurated the Christian Socialist movement in England. There too Adams was likely to meet the militant Radical, Frederic Harrison, who because of his distaste for "the shallow theology" of Hughes and Maurice had boldly struck out along the lines of Comte's "Scheme of Social Dynamics" as the chief English exponent of Comteism.[1] If these men impressed him as too visionary in their hopes for the good society, the same criticism could not be made of another good friend of the North, John Stuart Mill. Mill had warmly applauded Minister Adams's judicious handling of the "Trent" affair and in his article in *Fraser's Magazine* in February, 1862, he echoed the thesis of the Legation that the war was becoming more and more a matter involving the abolition of slavery. Even before Henry Adams met Mill, early in 1863, he had come wholly under the sway of his libertarian doctrines, and judged him "the ablest man in England." He passed "the intervals from official work," he said, "in studying de Tocqueville and John Stuart Mill, the two high priests of our faith."

Mill's new book *Considerations on Representative Government* systematized for him many ideas which had been household doctrine in the Adams family. In Berlin, for example, he had reminded his brother that however they approached political philosophy they both came down to the idea of universal

education as the source of New England's moral power. Horace
Mann, he said, had lived and died in that conviction; Goethe
too had shared it, believing that "his task was to educate his
countrymen and that all the Constitutions in the world
wouldn't help if the people weren't raised." Mill explained the
matter thus: "If we ask ourselves on what causes and conditions
good government in all its senses, from the humblest to the
most exalted, depends, we find that the principal of them, the
one which transcends all others, is the qualities of the human
beings composing the society over which the government is
exercised. [Hence] the most important point of excellence
which any form of government can possess is to promote the
virtue and intelligence of the people themselves." [2] These were
words which had belonged to John Adams as a representative
of the eighteenth-century Enlightenment. Liberty, said John
Adams, depends on the "intelligence and virtue of the people,"
and as he expounded the matter in his *Essay on the Canon and
the Feudal Law*, it is chiefly to be preserved by universal education.[3]

Through the years the statesmen of the family had elaborated
these intimations and provided them with new contexts; but the
spirit always remained the same. Henry was bound to read
Mill's authoritative prose, therefore, with a recurring thrill of
recognition. Renewed strength lay in Mill's reassurance that
the best government is government by the best. "A representative constitution is a means of bringing the general standard of
intelligence and honesty existing in the community, and the individual intellect and virtue of its wisest members, more directly to bear upon the government, and investing them with
greater influence in it than they would have under any other
mode of organization." But Mill cautioned him that though
men's convictions might be affected by "the united authority
of the instructed," the "natural tendency of representative government, as of modern civilization, is toward collective medi-

ocrity, and this tendency is increased by all reductions and extensions of the franchise." In this trend lay all the familiar horrors of the tyranny of the majority. The consequence in the United States, for example, was that "the highly cultivated members of the community . . . do not even offer themselves for Congress or the State Legislatures, so certain it is that they would have no chance of being returned." Henry could cite his father's success and yet in the same breath he would have to concede that Massachusetts merely proved the rule. And in a few years when Massachusetts again and again rejected his eldest brother's candidacy for public office he would learn that the rule was universal. Nonetheless at the moment the statement challenged him to resist the trend and show that a representative government could work. He could not help but believe, with Mill, that "one person with belief is a social power equal to ninety-nine who have only interests." [4]

The other "high priest" of his faith, Alexis de Tocqueville, ought not to have been a stranger to him. His grandfather John Quincy Adams had once known that friendly critic of democracy and his father had long been familiar with his work.[5] But Henry Adams now came to *Democracy in America* as a result of the widespread interest inspired by the Civil War, which friends and enemies of America regarded as a crucial test of that system of government. The entire pro-Northern coterie drew strength from de Tocqueville's appraisal of America's power of survival. The liberal M.P. Monckton Milnes had championed the "high priest" in a laudatory article. Henry Reeve, the influential editor of the *Edinburgh Review,* had once made a translation of *Democracy in America.* This translation was now being reissued in England and America. Nassau William Senior, the political economist, another intimate at the Legation, was in the midst of his nine-volume edition of de Tocqueville's works. John Stuart Mill's endorsement probably carried most weight. "The Montesquieu of our times," he called him, whose works "contain the first analytical inquiry into the influence of Democracy." Mill accepted his thesis that the movement toward democracy was irresistible and he also shared the Frenchman's view that the process was so filled with danger, as the individual sank into "greater and greater insignificance," that a new political science needed to be created for a new world.

Of course many of de Tocqueville's ideas had already entered Adams's thinking from other sources. Like the greater Guizot, de Tocqueville was a disciple of Montesquieu, whose political theories had been a commonplace of the Federalists from the time of John Adams's *Defence of the Constitutions of Government of the United States.* All of these men shared the fear of autocratic power and sought for means to restrain its operation. As an undergraduate, Henry had been taught by Guizot that the lesson of the French Revolution was "the danger, the evil, the insurmountable vice of absolute power." Now Mill underscored the lesson received from "universal tradition, grounded on universal experience, of men's being corrupted by power." De Tocqueville phrased this piece of immemorial wisdom in these words: "Unlimited power is in itself a bad and dangerous thing. Human beings are not competent to exercise it with discretion." Eventually these precepts in their progress through Adams's own writing became the harsh maxim of *The Education:* "Power is poison." [6]

Read at a time when he was becoming keenly alive to its implications, *Democracy in America* helped to catch and fix the random criticisms of America that had been cropping out in his letters—public and private—into a kind of catalogue of political imperfections. Where formerly he had written complacently of the American political system as of something perfected, yet destined by a law of necessary progress to indefinite material expansion, he now saw through de Tocqueville's eyes that "The organization and the establishment of democracy in Christendom is the great political problem of our times. The Americans, unquestionably, have not resolved this problem, but they furnish useful data to those who undertake to resolve it . . . If those nations whose social condition is democratic could remain free only while they inhabit uncultivated regions, we must despair of the future destiny of the human race." [8] The cogent analysis of the flaws in the American system fell in with the old warnings of his grandfather and father and revived their cankering doubts in him. "I doubt me much," he wrote, "whether the advance of years will increase my toleration of its faults." Still, de Tocqueville's text held open a perilous route of escape. If a great democratic revolution was still going on, it was susceptible of intelligent control. As in de Tocqueville's opinion, the true source of democracy in America lived in the New England character, Adams concluded that the rescue could be effected by the "New England element."

Here then was the mission that could give significance to literary and political action. "I have learned to think de Tocqueville my model," he announced to his brother, "and I study his life and works as the Gospel of my private religion." That religion promised him salvation on as formidable terms as any offered to his more theologically minded forebears; but he found comfort in the belief that "the great principle of democracy is still capable of rewarding a conscientious servant." A career was possible, if not as "a light to the nations," perhaps as a social and political reformer. In his own metaphor he saw "in the distance a vague and unsteady light in the direction towards which I needs must gravitate, so soon as the present disturbing influences are removed."

For such a role—still dimly conceived—much training would be needed. In a general way he indicated to Charles the nature of his self-imposed regimen: "I write and read, and read and write. Two years ago I began on history, our own time, and branched out upon Political Economy and J. S. Mill. Mr. Mill's works, thoroughly studied, led me to the examination of philosophy and the great French thinkers of our own time; they in turn passed me over to others whose very names are now known only as terms of reproach by the vulgar; the monarchist Hobbes, the atheist Spinoza and so on." The field of reading which he staked out was wide enough and it must have included the writings of such men as Thiers, Michelet, and Taine in France who were creating a great historical literature. Possibly he glanced at the theory of economic concentration propounded by the famous socialist historian, Louis Blanc, whom he met at a Legation dinner. The "great French thinkers" vaguely referred to may well have embraced Cousin, Proudhon, Comte, Littré, and possibly Caro, who was much concerned with science and materialism. Occasionally his reading carried him far out of his depth, as when he tackled Richard Owen's "Paleontology" [*Anatomy of the Vertebrates*]; "One idea in ten pages is the best I can collect," he confessed. "Geology is low comedy in comparison."

The questions of law raised by the "Trent" case and the "Alexandra" sent him scurrying for enlightenment in still another direction. "I am deep in international law and political economy," he reported to Charles early in 1863, "dodging from the one to the other." The materials in the Adams's London library being scanty, he wrote to Seward, at his father's direction: "This legation grievously wants a copy of the British admiralty reports; as well as of Grotius; Wheaton's History of International Law, and other books. But there seems to be no authority to buy them." By May 1 he was once more afield. "So I jump from International Law to our foreign history, and am led by that to study the philosophic standing of our republic, which brings me to reflection over the advance of the democratic principle in European civilization, and so I go on till some new question of law starts me again on the circle."

The "unfinished writing" of these early years in England

when he wrote merely to keep his hand in did not go wholly to waste. Among the salvageable materials undoubtedly were the notes for his article "The Declaration of Paris," which was finally published in his *Historical Essays* in 1891. This concerned the negotiation which he had so thoroughly misunderstood in 1861 when the British Ministry justly suspected Seward of baiting a diplomatic trap. It seems probable that he gathered together his impressions of that puzzling affair while it was still fresh in mind, as he had earlier summarized "The Great Secession Winter." Interestingly enough the final version of "The Declaration of Paris," thirty years after the event, loyally adhered to the view that Seward had acted in perfect good faith. Another rescued manuscript, on Captain John Smith, became, as we shall see, his first published article.

One composition had at least an official character. This was a formal report which he made on the great Trades' Union Meeting at St. James hall held on March 26, 1863.[9] Charles Francis Adams sent his son to cover the meeting, in place of the regular secretaries—Wilson and Moran. The gatherings, addressed by a number of Liberal leaders including Henry's friend John Bright, showed that the laboring classes though still without a vote were learning how to exert their influence in a manner that could not be safely disregarded by the Ministry. The upshot of the meeting may be gathered from one of Adams's concluding sentences: "It may therefore be considered as fairly and authoritatively announced that the class of skilled workmen in London, that is the leaders of the pure popular movement in England, have announced, by an act almost without precedent in their history, the principle that they make common cause with the Americans who are struggling for the restoration of the Union, and that all their power and influence shall be used on behalf of the North." His father, not dreaming that his official secretaries would resent this employment of his son, asked Moran to transcribe the covering dispatch. Furious at the fancied slight, Moran fumed in the privacy of his diary: "All law about unauthorized agents is to be disregarded and this son is to be pushed into prominence to feed his vanity and that of his father." Secretary Seward, to whom the Minister sent the report, praised it as "a very interesting analysis of the positions and relations of the parties concerned and a profound disquisition upon the import of the whole transaction." [10]

Through his association with the friends of Thomas Hughes, Adams became familiar with the political counterpart of the revolution in science. To his brother, whose concerns were so largely military and political, he spoke of the aspirations of the Reform movement. "You have no idea how thoughtful society is in Europe; even more so in some respects than in America, because there are practical hooks to hang thought on, like the church, education, poverty, and the suffrage, which are points all forgotten with us, but very much alive here and lead men far, when they once begin." In England the movement found expression during the closing years of his stay in such symposia as *Essays on Reform, Questions for a Reformed Parliament,* and *The Culture Demanded by Modern Life,* essays which embodied the progressive ideas discussed so often among his friends. His liberal friend Milnes (now Lord Houghton) contributed to the *Essays on Reform* the article entitled "On the Admission of the Working Classes as Part of our Social System." There were liberal articles also by Leslie Stephen and Adams's friend James Bryce. The latter, writing on "The Historical Aspect of Democracy," warned against the danger of class warfare and the evils that are bound to come from the growing concentration of wealth. Charles Francis Adams had uttered a similiar warning five years earlier: "The rich are growing richer, and conservatism gains rather than loses in its struggle for power." [11]

Frederic Harrison in his article attacked the senseless contradictions in England's foreign policy, and Meredith Townsend, whom Henry knew as editor of the *Spectator*, wrote on "The Poor." In the symposium on "Culture," men like Tyndall, Huxley, Paget, Faraday, Whewell, and Spencer indicated the direction which the education of the future must take. The keynote was struck by Youmans: "Deeper than all questions of Reconstruction, Suffrage, and Finance, is the question, What *kind* of culture shall the growing mind of the nation have?" It was the question that John Stuart Mill faced in his celebrated "Inaugural Address at St. Andrews" when he declared that the ultimate end of a university education was to fit men as "more effective combatants in the great fight which never ceases to rage between Good and Evil and more equal to coping with the ever new problems which the changing course of human nature and human society present to be resolved." For the young diplomat who would soon have to return to the United States to face a postwar world rotted through by political and commercial fraud that contest between "Good and Evil" would require that he bring to bear every scrap of wisdom which these writers offered.

Had Henry cut himself off from the Legation and tried for an independent literary career in London as an ally of his Radical friends, his small private income would have kept him in some comfort. But he could not bring himself to it. The uprooting process, though it progressed far and unfitted him ultimately for life in either Boston or Cambridge, could not overcome the pressure of family tradition toward public service. Mere literary success he had scorned from the beginning. Amidst the uncertainties of his position in London one thing at least was certain; he must somehow fit himself for a role in American life. When he passed twenty-five he acknowledged that the pretense of ultimately returning to Boston to study law in the office of Judge Horace Gray was no longer tenable. To the secretary of his college class he still wrote that he should be called "a student of law"; but he was more forthright with his brother. "Can a man of my general appearance pass five years in Europe and remain a candidate for the bar? We are both no longer able to protect ourselves with the convenient fiction of the law. Let us quit that now useless shelter, and steer if possible for whatever it may have been that once lay beyond it. Neither you nor I can ever do anything at the bar." A few years later when his friend Gaskell was admitted to practice, he told him: "I too once hoped to reach the proud distinction you have gained, but now that this hope has been permanently crushed, I hereby make over to you . . . all my right and title in any glory, gain or emolument whatsoever, which might have become mine in the pursuit of a legal career."

Yet what was he to do in the alternative? His mission might be clear, but precisely how to effectuate it? The men with whom he associated in England were going on to careers, each in his own sphere—law, politics, art, and literature. The inaction told on him. "It hardly seems consistent with self-respect in a man," he complained, "to turn his back upon all his friends and all his ambition, during such a crisis as this, only for the sake of conducting his mother and sister to the Opera, and a ride in Rotten Row." The demoralizing effect of "living as a Sybarite" troubled him. There was too much of the Puritan left to accept that as a proper career for an Adams. Sometimes the sense of ennui was so overpowering that he found himself "no longer able to enjoy America or Europe" and he wanted to "come home" if only to get a commission "on some cuss's staff."

At first, when it seemed improbable that he would remain in England more than a few years and prospects of an early victory seemed bright, he reverted to his old plan to settle in the West, ostensibly to practice law. The real objective, however, would be nothing less than the reformation of American society: "We want a national set of young men like ourselves or better, to start new influences not only in politics, but in literature, in law, in society, and throughout the whole social organ-

ism of the country—a national school of our own generation
. . . But with us, we should need at least six perfect geniuses
placed, or rather spotted over the country and all working together; whereas our generation as yet has not yet produced one
nor the promise of one." The proposal was strongly reminiscent
of Guizot's maxim which he had once read in college, that a
small group of "individual superiorities" would always rise to
power. With his eye on this grandiose scheme Adams was careful to "neglect no opportunities to conciliate" men like Evarts,
Seward, and Weed. He should like to have reached public figures further west, "but," as he put it, "the deuce is that there
are so few distinguished western men." His brother Charles,
who—for the time being at least—also acknowledged the futil-
ity of holding to their youthful plan to use the Law as a stepping stone to a joint career, held before him the true course
which they must pursue. More single-minded than Henry the
young officer on his way to a colonelcy did not fret over the
obstacles that lay before him. "All my natural inclinations tend
to a combination of literature and politics and always have.
I would be a philosophical statesman if I could, and a literary
politician if I must." In the complexities of London life, Henry,
torn between dreams of power and the fear that those dreams
would turn to "dust and ashes," found himself gravitating toward precisely the same objectives.

One factor perhaps more than any other delayed the reëm-
ployment of his literary talents. His will was subtly paralyzed
by the uncertainty of his position, and all his plans for a career
were affected by it. Not that his father would ever have dispensed with his services; in fact his job depended largely upon
his own filial determination to do the right thing. But always
there was the uncertainty of his father's tenure, which often
hung by a thread. Henry saw that he could have no plans so
long as his course was "tied to that of the Chief." In the early
days of their stay the family was unable to plan more than a

month ahead, and at frequent intervals Henry believed the termination of the Mission to be imminent. Seward's "Dispatch Number 10," the "Trent" affair, the "Alabama," the case of the Laird rams, each incident in turn seemed a peremptory notice to quit. As a consequence, he suffered "periodical returns of anxiety and despondency." A "season of calm" would give way to a "new period of squally weather." Sometimes when he forced himself to ignore the perpetual inner restlessness, he actually enjoyed all the "row and confusion immensely . . . as a spectacle and a study," with the same love of excitement experienced in Washington in 1860-61. Then, when the danger of foreign intervention faded he wrote that "existence floats along and time passes, thank God!" At such periods, work at the Legation dwindled and life became "merely a habit."

The Election of 1864 roughly jostled him out of his lethargy. The military news in July had been shattering: "Grant repulsed. Sherman repulsed. Hunter repulsed and in retreat. Gold, 250." Henry foresaw "the summary ejection of us gentlemen from our places next November, and the arrival of the Democratic party in power, pledged to peace at any price." These fluctuations of spirit provoked the usual impatient comment from Charles: "What the devil's up? What are you howling at? I never saw such a man . . . I see no sign that the American people and their policy are to be turned topsy-turvy just yet." The reproof did not quiet him for long, and a few days before the election he surrendered to his habitual "infirmity" of doubting "at the wrong time," and proceeded to canvass all the melancholy possibilities. "I am looking about with a sort of vague curiosity for the current which is to direct my course after I am blown aside by this one. If McClellan were elected, I do not know what the deuce I should do . . . But if Lincoln is elected by a mere majority of electors voting, not by a majority of the whole electoral college; if Grant fails to drive Lee out of Richmond; if the Chief is called to Washington to enter a cabinet with a species of anarchy in the North and no probability of an end of the war—then, indeed, I shall think the devil himself has got hold of us, and shall resign my soul to the inevitable . . . My present impression is that we are in considerable danger of all going to Hell together."

Charles was justifiably annoyed, for Henry's pessimism had a distinctly morbid cast.[12] It was both hereditary and constitutional. Recurring attacks of dyspepsia did not improve his self-acknowledged "bad temper." He had to rely a good deal upon pepsin and cod-liver oil, and when these failed he resorted to the waters at Baden-Baden. His pessimism seems also to have reflected a Calvinist survival in his temperament. On the death of a friend in 1862, he mused, "I have a kind of an idea that Stephen thought much as I do about life. He always seemed to me to take rather a contemptuous view of the world in general, and I rather like to imagine him . . . congratulating himself that at last he was through with all the *misères* of an existence that bored him." The mood could be evoked as well by the prospect of a friend's marriage. Thus, to his friend Major Higginson, he philosophized: "I only hope that your life won't be such an eternal swindle as most life is, and that having succeeded in getting a wife so much above the common run, you will succeed in leading an existence worth having." To Charles he once wryly declared, "I always was a good deal of a sceptic and speculator in theories and think precious small potatoes of man in general and myself in particular." Gaskell he solaced with a cynical jest: "I am led to infer that your hopes are disappointed. My dear boy, this world is a disappointment altogether. Let us quit it punctually next Sunday."

The succession of Federal defeats aggravated his recurring fits of despair until he was driven to gibber, "After having passed through all the intermediate phases of belief, I have come out a full-blown fatalist . . . The world grows just like a cabbage . . . The result will come when the time is ripe."

Sanity required him to acknowledge that this conception did very well only so long "as it had no occasion to regulate the relation between one man and another." Even Buckle's commonplaces on causation helped little. "The race will go on all right or wrong; either way it's the result of causes existing, but not within our reach." At such moments he needed the counsel of stoicism much more than his brother Charles, if only to face good-humoredly the fact that at twenty-seven he had grown "prematurely grey and bald." Greatly relieved by the reëlection of Lincoln, he was ready to believe it the beginning of "a new era of the movement of the world" and proof of "the capacity of men to develop their faculties in the mass."

Charles Wilson, the Secretary of the Legation, resigned during the campaign and returned to America. Seward elevated Moran to the vacant post and offered to appoint one of Minister Adams's sons. The position, which would have paid him $1500 annually, did not attract Henry, especially since his father planned to resign soon and he would be left at the mercy of Moran. He promptly declined the offer. "I wish nothing better than to be useful either directly or indirectly to the country in its difficulties," he wrote to Secretary Seward; "yet all that I have seen here in the course of the past three years has only strengthened my earlier belief that I should not be acting in the best interests either of the service of the Minister, or of myself, in accepting an official position in the Legation." [13] The decision was sufficiently prudent. Young Adams and Moran could hardly have worked together in double harness. The Minister and his son had grown indifferent to Moran's pretensions; whereas he construed almost every act of theirs as a deliberate insult.

In the summer of 1865 the Minister sent his resignation to Seward. Three months went by with no action taken. Again the likelihood of an early departure evaporated. By this time, Henry had become patient. "I have much doubt," he con-

fided to Charles, "whether we shall be released next spring, as
your papa pretends to expect . . . For my own part, I should
not object. I am doing as much for myself here as I should be
likely to do anywhere." The flurry produced by the election had
been succeeded by another period of calm. After peace came,
diplomatic activities declined to a minimum and Adams at last
turned seriously to writing, for want of other occupation. By
1867 he again speculated that "another effort will be made to
get out of our present situation." Six months later, the resignation of the Minister was finally accepted, to take effect on May
13, 1868, seven years to the day of his arrival in England.[14]

Henry left for the Continent on April 27, 1868, and returned
on June 26 to pay his last visit to the Legation. Both dates are
commemorated in Moran's diary. On the first he described his
last dinner at the Adamses. "Henry Adams," he wrote, "is losing tone. He is very conceited, and talks like a goose. But time
will soon settle him. He will ere long get a hard slap in the face
from fortune and that will sober him and give scope to the
good sense he really possesses." The second entry hints the
mellowing effect of Moran's promotion as chargé d'affaires.
"Henry B. Adams came and bade me good bye. He professed
friendship, but not very cordially and I don't think I shall ever
trouble him much. Still in the main he is a well-meaning and
kindly young man."

Path to Power

In November of 1865 Colonel Charles Francis Adams, Jr.,
visited London on his wedding trip. He had spent a two-month
furlough there in the spring of 1864. For a time, in the absence
of their parents, the two brothers ran the house in Portland
Place where, to Moran's annoyance, they luxuriated "in glory."
What hardheaded advice Charles gave his brother can easily
be surmised. Though already worn threadbare by correspond-

ence their plans for a postwar career needed to be worked out. The tall weather-tanned veteran of a half-dozen campaigns, who towered above his younger brother, argued to good effect. Reviewing the possibility of publishing some of the materials which Henry had been working over in history and finance, they fixed upon the two most promising items: a destructive criticism of the Captain John Smith-Pocahontas legend and a study of British finance at the close of the Napoleonic wars. For many months Charles had been preaching that "the management of our finances now seems to me not only the greatest but the most inviting field for usefulness which this country affords." Charles proposed to make a simultaneous bid for fame by working up an exposé of American railroads. When Charles Deane, editor of a recent edition of Captain John Smith's *A True Relation of Virginia,* came to dine on July 10, 1866, in the company of Harvard Professor Andrew Peabody, Henry had already begun his assignment.

Deane arrived in England in June, 1866, shortly after the appearance of his book. In the course of his conversation with Deane, Henry undertook to review that book and also Deane's edition of Wingfield's *A Discourse of Virginia,* the problem to be resolved being the veracity—or lack of veracity—of Captain John Smith with regard to the Pocahontas legend.

Adams had first become interested in the question in the spring of 1861, as a result of a conversation with his father's very good friend, John Gorham Palfrey, the eminent New England historian. Palfrey, then at work on his *History of New England* in the intervals of politics, had questioned the truth of the Pocahontas story. After Adams had settled down in England he resorted to the British Museum to "search after doubts." In October of 1861 reporting the results of his preliminary researches to Palfrey, he confessed that he was ready to "give it up." Captain Smith seemed an honest man. Palfrey called upon the expert knowledge of Charles Deane to set Adams right.

Chastened, the youthful historian went back to work, stimulated by the thought that the "Virginia aristocracy . . . will be utterly gravelled by it if it is successful." The "elaborate argument on the old pirate" was finished late in 1862, but was laid aside as unpublishable by its Horatian author. Palfrey did not forget the manuscript and after the war he reopened the subject in one of his letters to his young friend. Adams proposed to rewrite the whole manuscript and use Deane's book as a convenient "hook" on which to hang it.[1]

The choice of the Captain John Smith legend was a promising one for Adams's purpose, for it was as sacrosanct to Virginia as the Mayflower Compact to Massachusetts. An iconoclastic article on the subject might well create a sensation, as Palfrey had suggested, and get its author talked about as a fearless champion of truth in history. That Adams planned the article primarily as a political attack upon Virginia seems unlikely. Such animus would hardly be compatible with the humane position which he and his family were now taking on the Reconstruction question. Bad blood enough had existed between Massachusetts and Virginia and the Adamses were beginning to make political amends.[2] Nevertheless, Virginians were bound to take umbrage. Perhaps that price would not be excessive for public notice.

In his notes to the *Discourse of Virginia* Deane had suggested that Smith's inaccuracy might be shown by comparing the various accounts of his captivity in order to unearth the discrepancies. Palfrey in the original issue of his *History of New England* in 1858 had explained why he thought Smith's writings untrustworthy. With these materials before him, Henry set to work. The result pleased Palfrey and he undertook to place the article in the *North American Review*.

Adams pointed up for his readers the sensational implications of what he had written. "Stated in its widest bearings the question raised in this publication is upon the veracity of Cap-

tain John Smith . . . It will become necessary to reconsider
. . . the whole range of opinions which through him have been
grafted upon history . . . There are powerful social interests,
to say nothing of popular prejudices, greatly concerned in
maintaining their credit even at the present day." Having thus
thrown down the gage of battle he charged full tilt at the
patron saint of Virginia chivalry. The inconsistent passages
were shamelessly huddled together in parallel columns. With
true Brahminical righteousness he charged: "The statements
of the *Generall Historie,* if proved to be untrue, are falsehoods
of an effrontery seldom equalled in modern times." They deserved to be exposed exhaustively. Therewith step by step, he
heaped up his evidence with legalistic thoroughness.

What especially irritated Southern readers of his day, as it
continues to upset the neo-Confederate historians of the present, was the patronizing tone that colored his criticism. "Each
one of these four stories is more or less inconsistent with all
the others; but this one of 1622 is, we are sorry to say it, more
certainly mendacious than any of the rest. Read it in whatever
light we please, it is creditable neither to Smith's veracity nor
to his sense of honor." To account for the doughty captain's dis-
honesty he advanced the hypothesis that Smith invented the
story of his rescue by Pocahontas in order to secure public employment, an ungenerous judgment in view of his own reasons
for writing the article.

Adams asked Palfrey to forward the article to Charles Eliot
Norton, editor of the *North American Review,* to whom he enclosed a letter, referring to it as the one "which Dr. Palfrey has
been so kind as to take under his protection." There was a
certain appropriateness about the choice of the *North American* for his first flight as it had been the organ of his forebears
almost from its founding in 1815. His father had himself contributed at least seventeen articles during the days when he
was closely associated with Palfrey. As Palfrey had been the

proprietor and editor from 1835 to 1843, and an important contributor thereafter, his recommendation carried weight with the new editor. Toward the end of 1866 Henry was able to report to Deane, who had already returned to Boston, that "Mr. Norton will put us into the January number." Delighted with the article Deane thought it an admirable presentation. Norton, similarly impressed, ran it as the lead article, after improving it by a few minor editorial corrections.[3]

The thirty-page study did create its little sensation. Although it was unsigned, as was the practice at the moment, the identity of the author was soon common property. The *Nation* promptly played up the sectional implications: "The name of Massachusetts . . . will henceforth be more bitterly execrated than ever . . . We doubt if Mr. Adams's arguments . . . can be so much as shaken." In the South the aspersion upon the womanhood of Virginia was particularly resented. The editor of the *Southern Review,* after pondering the insult for two years, during which the *Nation* and "many other journals" had "spread abroad the doubts set forth in the *Notes* of Deane," denounced the "historians who deal in hints, innuendos, and dark insinuations . . . especially when their oblique methods affect the character of a celebrated woman." What more could be expected from "two knights of New England chivalry" who maliciously perverted history because Captain Smith had "labored in a different latitude" and because Pocahontas had not "been born on the barren soil of New England"?[4]

Before dispatching the Captain Smith article, Adams rewrote the manuscript of an article on British governmental finance. It accompanied the Smith article. The covering letter to Editor Norton was a model of diffidence. "My absence from America has been so long that I know scarcely how or where to turn to get a public hearing on any large and rather difficult subject, else I should not try your patience by sending you two manuscripts at one time. If I were not still more ashamed to continue the process, I should express a wish to be allowed to offer you a third Essay, on the Suspension of Specie Payments in England from 1797 to 1819. If however you have the patience to wade through the two heavy articles which I now send you, and to accept one of them, it is all that I can reasonably wish." Norton accepted both manuscripts, printing the first in the January 1867 issue, one month before Charles's first article on the railroads; and the second in the April issue.

Currency questions had occupied a large place in the wartime correspondence of the Adamses, trenching as they did on traditional family doctrines. Seventy years earlier John Quincy Adams had condemned legal tender acts as "among those measures which would give the sanction of law to private fraud and villainy." In 1837, a year before Henry was born, his father, shocked by the disastrous panic which followed Jackson's abolition of the United States Bank, had denounced the evils of an irredeemable currency and had admonished the country to "work back as soon as possible to the true principles of banking—shortened credits and a rigid settlement in coin." Adam Smith confirmed him in the belief that tampering with the currency was a "juggling trick." John Stuart Mill called it an "intolerable evil" and "a form of robbery." Hence, when in the early days of the Civil War Secretary Chase's desperate expedients to raise funds ended in the suspension of specie payments, Henry Adams felt as if he were witness to the judgment of Heaven. "Financially we are dished." [5]

Henry Adams was not the first American writer to turn to England's monetary experience during and after the Napoleonic wars for instructive parallels. Early in 1862, for example, about a week after his unexpectedly final contribution to the New York *Times,* a long article appeared in that newspaper called "War Finance in England." It discussed the Bank Suspension of 1797 and pointed out that the "development of financial principles which were made in England during this period are of great importance to us at the present moment." The proponents of a fiat currency, he said, could be expected to seize upon the English experiment as a justification. The parallel he showed was a dangerous one. It may be assumed that that article was read with especial care by the sound money men of the American Legation.

The first half of Adams's "British Finance in 1816" reviewed "the financial difficulties with which England at the close of the war had to struggle so far as they related to the revenue." The intricacies of the subject defied simplification. He cautioned his readers: "This account is difficult to understand, and may contain errors; but there can be no possible doubt that a clear and positive statement would be quite unworthy of belief." Among the "financial difficulties" he singled out for special attack the exactions of the protective tariff for he had long since been won over to free trade by Mill and those two leaders of the Manchester school, Cobden and Bright. "Protection coiled like a tangled cord around and over and through every portion of British finance . . . Everything was protected. Every petty interest of the country had its rag of protection,— not merely against the genius or activity or superior circumstances of a foreign rival, but against allied branches of industry at home. Tiles complained if slates were untaxed. Wool was jealous of cotton . . . For an entire century, down to 1831, the British people were condemned to drink the vintage of Portugal, in order to protect both countries against the superior attractions of French wine."

The prose rhythms of the essay show that the measured invective of Junius and Burke still enchanted him. He delighted in the exquisite hammer blow of the periodic sentence with its scrupulously balanced parallels: "A protective system so logically perfect, so rigid in its obedience to a single purpose, so universal in its scope, so minute in its application, contains elements almost of grandeur." He knew the impressiveness to be gained by symmetrical massing of detail. "The student, who has labored month after month to comprehend it [the finances], turns away at last with despair, and abandons the attempt as not worth, even if successful, a tenth part of the mental effort that is required to fathom it. There was no beginning, no middle, and no end. There was a bottomless ocean of accounts, without a system, without connection, and without result . . . The mysteries of the Exchequer were portentous. Few Englishmen, even then, had any conception what functions were exercised by such mediaeval creations as the Department of the Pells, the Pipe Office, and the Tally Court, the calm abodes of happy sinecurists, whose duties of inconceivable clumsiness had long been assumed by the Bank of England and other offices, but who were still permitted to perform by deputy an imaginary routine." The laborious writing and rewriting that Adams had been inflicting upon himself was now bearing fruit and although the style was still excessively judicial and stately, the cadenced sentences must have pleased the select audience which he wished to reach.

In the second half of "British Finance" he turned to the question of the currency as "another evil which lay behind them all," but the few paragraphs which he devoted to the subject did little to lighten the general obscurity. He recognized this difficulty and prepared the way for his next article by stating that "The history of the process by which Great Britain succeeded at length in restoring its original standard of value is one so important, so instructive, but also so complicated and disputed, that it cannot be dealt with in a few sentences. It requires an entire chapter in itself." Much more engaging were a half-dozen pages of literary digression in which he displayed a marked advance in his command of portraiture. In his sketches of Vansittart, Huskisson, and other of Lord Liverpool's "followers," he managed however to be more striking than accurate, by avoiding all pedantic half-lights. "Mr. Vansit-

tart was not a man gifted with the blind toryism of Lord
Sidmouth, or the narrow-minded perverseness of Lord Eldon.
He was simply a thoroughly incompetent mind. It would
scarcely be worth the while to dwell upon his qualities at any
length, if it were not that he was almost a perfect representative of the old school of financiers . . . a school . . . which
lent the influence of narrow minds to encourage and aggravate
the mistakes of great ones."

To the American reader who might overlook in this account
of financial confusion a tract for the times, he alluded to the
"interest which belongs to it of furnishing instruction to other
nations which are placed in circumstances more or less similar."
The hint could not have been more elaborately indirect nor
more loftily disinterested. It was his first bid for acceptance
as a philosophical statesman. Beneath the philosopher's toga,
however, the rhetorical "stilts" of his youthful style were still
occasionally visible.

The care with which he revised the article proved time well
spent. Norton made very few corrections and Adams acknowledged them with his usual deference: "Such omissions as you
have heretofore made in my articles, so far as I noticed them,
have been decided improvements." The corrections amounted
to no more than the rearranging of two or three sentences, the
alteration of a word or two, and the omission of a single piece
of high-flown rhetoric: "The modern student who wanders
painfully among these ruins of a work of so much genius, and
who is condemned to find a path where no path exists, hurls
anathemas at the memory of Mr. Pitt and stands aghast at the
incompetency of his successors."

By February 1867 the second financial article, on the Bank
of England Restriction, began to take form. It turned out to
be "an elaborate and unexpectedly difficult essay" and the task
dragged itself out, but by the end of June it was finished.

The new article made a detailed study, complete with statistical tables, of the suspension of specie payments in 1797. Adams showed that the problems involved in the suspension in 1797 and the resumption in 1821 were unreal if evaluated in isolation. Aiming at being "more philosophical and more exact," he demonstrated that the statutes which Parliament adopted, treating the Bank issues as if they were independent of a complex price and credit structure, had little consequence for good or evil. "The only effect of the long suspension was to breed a race of economists who attributed an entirely, undue degree of power to mere currency, and who for years to come delayed a larger and more philosophical study of the subject, by their futile experiments upon paper money." The burden of the article was that no proper comparison could be drawn between the British and the present American experience because "whatever action may have been caused by the Restriction was upon credit in the first place, and not upon the currency"; moreover the depreciation in England was far less violent than in the case of the American greenbacks. He thus could reassure the banking interests that the resumption of specie payments, which would be so advantageous to them as a creditor group, might be immediately attempted without necessarily precipitating the evils feared by the debtor classes.

The two articles constituted a remarkable tour de force for an amateur economist.[6] Of the first the *Nation* respectfully—and briefly—observed " 'British Finance' is a paper of whose value we can hardly speak. It is made quite interesting to the general reader." As for the second article, the *Nation's* reviewer concluded that its analysis of the English experience with inconvertible currency offered no comfort to the United States. Adams's own opinion of the article expressed his usual self-mistrust. Fearing the essay was dull, he requested that Norton "not publish the thing at all, if it is not readable." As for its excessive length, he apologized that he had "crowded books into sentences and years into lines . . . But I am happy to pass the scissors over to you." Norton, realizing the merits of the article, declined the scissors and printed it practically without correction in the *North American* for October, 1867.

The letters to Norton of February 28 and June 28 of that year remind us that Adams had not yet got over the fright caused by the roasting given him in the British press five years before. In the earlier letter he asked that his name not be added to the article on "British Finance," the practice of identifying the authors having just begun. "I have suffered so much from publicity that I prefer over-caution." In the later letter he nervously repeated his request. "I have this very season undergone two whole columns in the *Pall Mall.*[7] In so concentrated a society as that of London, and in so exposed a position as I am occupying here, this process is annoying. But my objection applies merely to newspaper notoriety, and I am very far from objecting to any private avowal of authorship."

Having published three weighty articles in the leading American quarterly, he would have been justified in congratulating himself as a coming writer. His opinions commanded respect. He had made an enviable beginning as a publicist. Yet the effort seems to have exhausted his nervous energy, and he was swept with a sense of revulsion. Into a bewildering letter to Charles, he tumbled all of his dissatisfaction. "The triumph of earning $240 in paper in one year does not satisfy my ambition. John is a political genius; let him follow the family bent. You are a lawyer and with a few years' patience will be the richest and most respectable of us all. I claim my right to part company with you both. I never will make a speech, never run for office, never belong to a party. I am going to plunge under the stream. For years you will hear nothing of any publication of mine—perhaps never, who knows. I do not mean to tie myself to anything, but I do mean to make it impossible for myself to follow the family go-cart."

Still stranger moods took possession of him as, for example, when he suggested to Gaskell that they stir up each other's acquaintances to say ill-natured things about each other. "I would do anything to experience new sensations, even disagreeable ones, and a good, spiteful, vicious attack is such a tonic!" In spite of these vagaries he did not lose sight of his plan of "practical training" for a newspaper career. In April 1867 he asked his brother to subscribe to the newly founded *Nation* for there seemed "to be merit about it . . . I wish also that you would send out to me anything that appears worth noticing in American literature, anything new, I mean, which seems to you capable of standing criticism. If I write here, I shall write what I think, and not be soft on people's corns."

His energetic reconnaissance in the fields of history and finance completed, Henry was now ready to develop a suggestion made to him by Sir Charles Lyell. Early in 1867 Lyell had happened to discuss with him the possibility of getting the recently published Tenth Edition of his *Principles of Geology* "properly noticed" in America. Adams brashly volunteered to do the job himself. An allusion in one of his letters to Charles Milnes Gaskell hints that he proposed to do some field work at Liège by way of preparation. "It is easy enough to write on the subject without any acquaintance with it," he quipped, "but a dash of truth would add a certain base to the dish." The interest in geology inspired by Agassiz had been strengthened by years of friendly intercourse with Sir Charles. That interest soon grew into a passion for geologizing along Wenlock Edge with the intimates of Gaskell's circle. But to review a significantly new edition of the classic of that science he needed more than a dilettante's information. As a result the work progressed slowly and was subject to many interruptions. Really serious application had to wait until "The Bank of England Restriction of 1797" was out of the way. In August the great Paris Exposition distracted him with its "chaos confounded." Then as family escort he went off to Baden-Baden to take the cure once more and to amuse himself occasionally by risking a five franc note at roulette.

Still he kept mulling over the problems to be solved. The problem of style was the most insistent. He wrote to Gaskell: "I hope you keep on writing, but after all, writing is only half the art; the other being erasure. No one can make real progress that doesn't practice the latter as vigorously as the first. I have done it so effectually as to have expunged all my last thing." This "last thing" one surmises was an initial draft of the Lyell review. Back in London he set to work again but again it was the labor of Sisyphus. In March, 1868, just before leaving for a tour of the continent with the family, he sent a submissive prospectus to Norton, apparently to forestall possible objection. "I ought perhaps to add that I shall try to express more valuable opinions than my own, though I don't wish to be controversial. As it is long since I have had the pleasure of talking with Mr. Lowell [Norton's coeditor], and neither your opinions nor his are very well known to me, I would rather run no risk of offering to you anything which might seem not conservative enough for your united tastes. Therefore if you are afraid of Sir Charles and Darwin, and prefer to adhere frankly to Mr. Agassiz, you have but to say so, and I am dumb. My own leaning, though not strong, is still towards them, and therefore I should be excluded from even the most modest summing up in the *Atlantic*, I suppose. It is not likely that I should handle the controversy vigorously—the essay would rather be an historical one—but I should have to touch it." Even in Paris he continued to wrestle with the manuscript, humorously complaining: "Thus far I have passed my days and nights in my room geologizing." Still the article would not march. Then a new complication arose. His father's resignation was at last accepted. The imminence of departure for America discouraged any further work. Not until he reached Quincy did he return to the formidable task, finally completing the paper late in August, 1868, a year and a half after he undertook the assignment.

The thirty-five page review appeared in the October *North American* above the signature of "Henry Brooks Adams" and thus avowedly for "reputation." It had for company articles by Leslie Stephen and Henry James. The conservative Boston *Transcript* graciously alluded to it as an "instructive paper." As Adams's first literary venture into science, the essay represented an achievement as great in its way perhaps as his grandfather's uninstructed investigation of the science of weights and measures. Norton may have advised him to take a conservative view; but if he did, that advice was unnecessary, for Adams's treatment of his two protagonists—Lyell and Agassiz —indicates that at heart he was on the conservative side.

The review of the *Principles of Geology* limited itself to a consideration of "some of the most striking changes of view, which make the tenth edition almost a new book." These included, first, Lyell's explanation of the operation of the climatic element in geology, and second, his acceptance, through Darwin, of "the theory of progressive development of species." To explain the variation in climate between "Miocene warmth and glacial frost" Lyell argued that there had occurred an oscillation in the distribution of land masses about the poles and the equator. Adams objected to this explanation with such penetrating insight that later the questioned chapters were recast.[8] Lyell supplemented his hypothesis with that of James Croll who also rejected Agassiz's catastrophe theory of glaciation. Croll suggested that the alterations in terrestrial temperature could be explained by the eccentricity of the earth's orbit. Adams relentlessly pursued this second explanation through the "difficult reasoning and calculations" only to conclude that the theory did not "accord with the facts. The present eccentricity of the earth's orbit is very slight, so that, after admitting the glacial period to have been brought about in the manner described, a still greater difficulty arises in attempting to account for the antecedent warm period, and for an average degree of heat greater than can be obtained by calculation. We do not understand how this objection can be met." He had placed his finger on one of the most vexing puzzles of geology. The objection has not yet been met. Pleistocene glaciation according to a recent text continues to be "one of the least understood and most disputed problems of geology and meteorology." [9]

Adams next examined the process by which Sir Charles had come to accept Darwin's hypothesis. He proceeded to summarize the "arguments in regard to the variability of species," as illustrated in Darwin's account of the rock pigeon and in Lyell's description of the flora and fauna of the Madeiras and the Canaries. These arguments he said led "Sir Charles Lyell . . . to adopt opinions which many excellent men consider revolting." However, Adams continued, "a more tangible objection than this repulsiveness, and one more likely to influence scientific men is, that Mr. Darwin has, after all, announced only a theory, supported, it is true, by the greatest ingenuity of reasoning and fertility of experiment, but in its nature incapable of proof."

The alternative to this theory was of course the "hypothesis of special creation" implying "a vital force which always strives to realize" the prior conceptions of a divine mind. The physiologist, he conceded, might be obliged to choose between the development theory of Lamarck and Darwin on the one hand and the theory of special creation of von Baer and Agassiz on the other; but as for the public at large it was not equally bound, for "extremely little is as yet known with certainty on the subject, too little to warrant any unscientific person in becoming a partisan of either opinion, and far too little to justify one party in announcing dogmatic conclusions, or in excommunicating its opponents." This suspension of judgment was especially needed because "neither of the two parties which are in such eager dispute has as yet fairly met the problem which is at the bottom of every one's thought. How is it that the rock-pigeon has bequeathed . . . qualities which it did not possess itself . . . ? Here Mr. Darwin's theory explains nothing; it merely records a fact. We here step beyond the range of science, and begin a hopeless attempt to struggle with first causes." [10]

The elaborate argument rewards careful reading because underneath its air of Olympian detachment one sees that in his own mind Henry continued to subordinate science to conventional metaphysics as completely as did Agassiz himself. And his patronizing treatment of Lyell and his ideas is precisely what one would expect from a philosophical idealist who, speculating about first causes and the unity of things, rejects the more modest ambitions of the scientist. He had vigorously assailed two of the main theories of Lyell's text, yet when he touched on the hypothesis of special creation he suddenly seemed to remember that he had promised not to "handle the controversy vigorously" and turned away from it with a sally at Lyell's expense. "We have no intention of questioning the merits of this hypothesis; but we can easily imagine that to Sir Charles Lyell it must have appeared inadmissible, since Sir Charles's whole strength lies in the direction of Realism, while this is essentially an idealistic conception, such as few English minds could grasp at all. To fancy, Sir Charles Lyell, of all Englishmen, in the act of grappling the notion of a type controlling an organism, would be an association of ideas so incongruous as to approach annoyingly near the laughable."

Sir Charles's reluctance to be swept off his feet by Agassiz's glacial theory also seemed a cause of reproach, yet Adams implied no corresponding fault in Agassiz's far more stubborn hostility to Darwin's theory. "We have already mentioned the unwillingness shown by Lyell to accept the doctrine of great [i.e. catastrophic] climatic changes, when Professor Agassiz, stepping so boldly out of his own strict sphere of science, forced upon geology his celebrated glacial theory." He felt that Lyell's "geographical theory does not have quite so large and liberal a character as we might wish."

It was a strangely muddled comment, for his master Agassiz was even then writing to an English friend, "My recent studies have made me more adverse than ever to the new scientific doctrines which are flourishing in England . . . I trust to out-live this mania also." Perhaps Adams had in mind one of the cautions uttered by Asa Gray in the *Atlantic:* "The English mind is prone to positivism and kindred forms of materialistic philosophy, and we must expect the derivative theory to be taken up in that interest." Gray was content to believe however that the theory did not "affect the argument for design," as he was less intensely religious than Agassiz.[11]

Adams dealt with his former professor Agassiz in an entirely different spirit. Not a word escaped him concerning the de-structive criticisms which had long been made of the theory of catastrophism. Agassiz's curious mélange of theology and sci-ence did not astonish him. On the contrary, he believed that the "glacial theory" was "certainly the most brilliant discovery of the last half-century." To him Agassiz's eloquent description of prehistoric glaciation as a sudden apparition destroying all life at the end of each epoch was satisfyingly large and liberal and it was expressed in a sufficiently imaginative style. His theory had grandeur and finality.

A few months after the article appeared in the October 1868 *North American* Adams suffered a moment of self-abasement. "My article on Lyell humiliated me," he admitted to his brother Charles. "It was so neatly put together and not an original idea in it." However, the indulgent Lyell wrote him "a very handsome letter" in which he described it as "the most original he has yet seen on his new edition, and the only one which has called due attention to what is new in it." The *Nation* gave him a reassuring pat on the back, declaring that the author of the review "talks learnedly and, to our mind, sensibly." [12] Norton paid him $100 for it.

With this article Adams's position as a permanent contributor to the *North American Review* was assured. He had faithfully followed his father's precept and example and had not wasted his talent in the sort of light literature cultivated by the *Atlantic Monthly*. He had demonstrated a respectable ability in dealing with problems of history, finance, and science. In a sense he had passed his self-imposed examination for the job of "philosophical statesman." True enough there was little money in such writing, even if it should lead to a New York newspaper, but that consideration would not now deter him. He was playing for higher stakes. More than ever the country had desperate need for a national school of bright young men of his own generation. He had begun to train himself for a national career and his writings had demonstrated his right to be heard in America.

Chapter Six

A Young Reformer of Thirty
1868–1870

A Power in the Land

U NKNOWN TO THE ADAMSES on the stifling evening of July 7, 1868, as they disembarked from the "China" in New York harbor, was the fact that their eldest son, John Quincy Adams II, had just received the compliment of one vote for the Democratic presidential nomination. Cast by a South Carolinian at the Tammany Hall Convention, the vote gratefully acknowledged the political apostasy of the grandson of their old adversary, ex-President John Quincy Adams.[1] It was also a bad omen of what was in store for the family as a power in Massachusetts politics. Twenty years earlier conscience had driven them to revolt against the cotton leadership of the Whig party. The revolt which they had helped direct had succeeded too well, for Conscience had become Radical Republicanism. Throughout the war the family clung to the principles of the moderate faction, hoping to fight the issue out within the Republican party. Only John Quincy rebelled. Unable to stomach the divisive tactics of the Radicals he journeyed to Philadelphia in August 1866 as a delegate-at-large to the National Union convention, which had been called to rally support for President Johnson's policies. The gesture was widely taken as an official act of the family and the Boston *Advertiser* demanded ostracism. Protected by Secretary Seward, the Minister had so far escaped reprisals. But the opportunism of Seward proved

no match for the fanaticism of the Radicals. And at the head
of the Radicals stood Charles Sumner.

The first blow fell early in 1867. The Senate rejected the appointment of John Quincy II to the Customs House at Boston.
Shocked by this illustration of "extreme partisanship" the
Springfield *Republican* protested that "it was done at the in-
sistence of our Massachusetts senators, and we suppose to
punish Mr. Adams and his father, our Minister in England, for
presuming to differ from them as to questions of reconstruction." Equally punitive was the bill introduced by Sumner,
outlawing court dress for American diplomats. Obviously aimed
at Minister Adams this piece of senatorial spite became law.
Disgusted by the mean-spirited actions of the leaders of his
own party, John Quincy went over to the Democrats without
further delay. The quarrel with Sumner had not been patched
up in spite of Henry's flattery in the *Advertiser* and his placa-
tory report on the Massachusetts appointments. Late in the war
Dr. Palfrey helped Charles Francis, Jr., resume social relations
with Sumner, but Sumner, as Charles said, "did *not* press me
with any inquiries about the Minister or his family." He was
not thus to be appeased. He did not join in the celebration to
honor the return of Minister Adams, and an obliging friend
consoled him with a report that "the hall was only half full." [2]

The Adamses were now hopelessly alienated from their Ultra
colleagues in the Republican party. In the course of the war
the Moderate group had been successfully liquidated and at
its close the Adamses were left without a following in their
own party. Their program of reconstruction had from the beginning followed the intimations of Charles's articles in the
Atlantic and the *Independent:* the South would have to regenerate itself. Henry Adams and his brother debated at great
length the methods by which, at the close of hostilities, the
South could be restored to the Union with least shock to the

Constitutional system. Slavery, they agreed, must ultimately be abolished; its immediate abolition might even be a military necessity; but no measures were justified which might impair the moral supremacy of the Constitution. In the main they cordially approved the moderation of Lincoln's and Johnson's theory of reconstruction, although their father sharply questioned the legality of Lincoln's methods. "The President's proclamation as well as most of the plans of reconstruction of the state authorities which were offered in Congress seem to me," he told Dana, "to rest upon a mistaken idea of the powers vested by the Constitution. As President, Mr. Lincoln unquestionably had no power to emancipate a single slave. Neither had Congress the smallest right in my mind to meddle with the reconstruction of a single state." [3] He therefore approved President Johnson's veto of the Freedman's Bureau Bill believing that if the South could not be retained by kindness it must be allowed to go its own way. He likewise opposed giving the vote to the Negroes. To the secretary of the legation, Moran, these sentiments seemed "rank secession." As the minister made no effort to conceal his opinions, his former colleagues in Massachusetts were filled with alarm. After his son bolted the party Richard Henry Dana wrote posthaste to London to sound him out, "tormented by the fear," according to Henry, "that our chief is going over to the Democrats." His father's reassurances did not end speculation.

When Congressional reconstruction became an accomplished fact, Henry could hardly contain himself. "I tell you frankly," he told Charles, "that when I think of the legislation since last year, my blood boils." He therefore admired their brother John for what he had done and impatiently rejected as "a piece of self-delusion" Charles's suggestion that it would have been more expedient to fight the issue out "within the party." Looking beyond the presidential election of 1868 Henry prophesied that once Grant was elected there would be "a new division of parties" and that the Adams brothers would not be "in the Republican ranks."

In November of 1867 the eldest brother, eagerly welcomed by the Democrats, was nominated for the governorship. He lost the election, but the 65,000 votes cast for him seemed to Henry Adams a suitable reprimand to Senators Sumner and Wilson "for the insult they thought proper to put upon us last summer in respect to the custom house." John was again nominated in 1868 with the blessing of President Andrew Johnson to whom he had written: "My grandfather used to predict that when the great slavery struggle which he saw impending closed, there must come a great constitutional party or anarchy, and today it seems to me high time for calm and patriotic men to be gathering around the organic law." In his letter of acceptance, he indicated that he would not compromise his family's conservative economic principles. To the greenback advocates among the Democrats, he announced, "I believe in hard money." [4]

The election that carried Grant to the presidency also saw Sumner returned to the Senate to continue his leadership of the vindictive majority. John Quincy Adams was of course again defeated. The Adamses did not believe, however, that the main fight was lost for they all concurred in their faith in General Grant. Back in 1864 Charles thought him "a very extraordinary man . . . of the most exquisite tact and judgment" with "all the simplicity of a very great man." And during the height of the campaign, John Quincy Adams told a South Carolina audience that even though he was an opponent of Grant he believed him to be "an upright, honorable man who will try, if elected, to do his best, not for a party only, but for the whole people of the country." Their father according to the Hartford *Post* also expressed himself "in private conversation with intimate friends . . . as strongly in favor of Grant and Colfax," but he carefully avoided any public statements. Shortly after Grant's election Seward loyally urged that Charles Francis Adams should succeed him as Secretary of State. The story that Grant might appoint the distinguished ex-Minister was soon out. Secretary of the Navy Gideon Welles jotted in his diary: "Sumner is much disturbed with this rumor." Henry, believing that the "first Cabinet will be a failure" disliked the scheme. The move collapsed, and Henry and Charles Sumner were both greatly relieved, but for quite different reasons. On the other hand, the faithful New York *Times* lamented: "It is among the most prominent infelicities of our practical politics, that Mr. Adams was set aside." [5]

This then was the political situation of the family shortly after Henry Adams left Boston on October 12, 1868, at the age of 30, to seek his fortune in Washington. He carried with him "a short line of introduction" from the well-known Boston economist Edward Atkinson to Edwin L. Godkin, editor of the *Nation.* While in New York, en route to the Capital, he made arrangements to write for the *Nation,* the *Post,* and possibly the *Sun.* Why he did not communicate with his old editor, Henry Raymond of the *Times,* is still an unexplained puzzle. Raymond contrary to the statement of *The Education* was still very much alive, his sudden death not occurring until nine months later. In the course of his journey Henry was joined by his friend William Evarts, the new United States attorney general, but any hope that this encounter gave him of working from within the Administration, was soon dispelled by Grant's victory in November over the less colorful Horatio Seymour.

Reaching Washington on October 22, he was taken in tow by the Attorney General who presented him to Andrew Johnson. The President, "grave and cordial," gave him "a little lecture on Constitutional law." Seward "was also cordial," and before long Adams was in the good graces of Secretary of the Treasury McCulloch and Secretary of War Schofield. Grant's shadow lay upon the Cabinet, but young Adams felt no insecurity for early in November he wrote Gaskell: "The great step is taken and here I am, settled for years, and perhaps for life." After the expenses of London, he found himself able to "live comfortably and cheaply," well within his private income as eked out by his earnings as a journalist. It was comforting to him to reflect that "the members of Congress and the Cabinet have only about twice my income or less." [6]

While waiting for the Fortieth Congress to reconvene on December 7, he began to study the subject which he had made his specialty in England—government finance in all of its ramifactions, including the greenback currency, the redemption of government bonds, the modification of excise taxes, the adjustment of the tariff, and the reduction of federal expenditures. As his brother had once said it was "the most inviting field for usefulness." Of course Southern reconstruction still bulked large on the political horizon, but it promised no career for a person of his anti-radical views. The Thirteenth and Fourteenth Amendments had already been ratified, and the Fifteenth, guaranteeing the vote to the Negro, was about to be proposed. If the recent impeachment trial of Andrew Johnson taught anything, it taught that Congressional reconstruction must run its course. No mere journalist dared stand in the way.

For textbooks in finance he used Secretary McCulloch's Annual Reports and the brilliant compilations of David A. Wells, the Special Commissioner of the Revenue. Wells had been lecturing Congress "like a schoolmaster," there being, in Gideon Welles's phrase, "not a single financial mind in Congress." But if Congressmen were ignorant, editors were no better, being "perfect blockheads on the subject of currency." Wells saw in Adams a chance to train a fellow schoolmaster. His protégé could hardly have found a more brilliant instructor. Wells, with whom Henry had become acquainted in London, perfectly exemplified the supremely capable and disinterested public servant to whom he could look for guidance. Wells's new report on the revenue delighted him. "I needed it," he wrote Wells. "The feeling of disgust which was beginning to get control of me after two months study of our system and the Congressional Globe, was becoming too strong for comfort. This report of yours is the first statesmanlike expression of policy I have seen." He had sounded out his friend Henry Reeve, editor of the *Edinburgh Review*, on the marketability of an article on American finance as early as November 8. By that time his co-worker Charles was already well along on his railroad series, having completed his first exposé of the Erie Railroad management for the October 1868 *American Law Review*. Now he was busy with an article on "Railroad Inflation" for the January *North American.*[7]

Not having Charles's leisure or single-mindedness, Henry proceeded more slowly with his article. The opening of Congress brought to Washington not only legislators but a swarm of newspapermen and lobbyists. From the beginning he attached himself to militant critics like Charles Nordhoff, the correspondent of the New York *Post;* Murat Halstead, editor of the *Cincinnati-Commercial;* Henry Watterson, editor of the Louisville *Courier-Journal;* and the unquenchable Sam Bowles, editor of the Springfield *Republican.* He also formed close ties with Wells's able collaborator, Francis A. Walker, chief of the Bureau of Statistics. This coterie of "working practical reformers," as the *Nation* called them, made itself felt in Congress, chiefly through Representatives James A. Garfield and Thomas Jenckes. When, for example, Representative Benjamin Butler spoke in favor of Greenbacks, Adams informed his brother: "Garfield, Wells, Walker and I have held his inquest, and Garfield will score him in the House." For an "insignificant cuss" as he called himself Henry Adams was making progress behind the scenes.

His first contribution to the crusading *Nation,* a report on the Legal Tender Cases, grew out of his friendship with Attorney-General Evarts. In prolonged conversations with Evarts during November 1868, he talked as Devil's Advocate against the constitutionality of the wartime greenback bill. *Hepburn vs. Griswold* and *Bronson vs. Rodes* came before the Supreme Court for hearing on December 8 with Evarts and Benjamin Curtis on the government's side in support of the constitutionality of the Act. "After listening for three entire days to the arguments," Adams hurried off his appraisal of the trial to the *Nation,* which published it on December 17.[8] As a hard money man and a strict constructionist, he expressed "a certain feeling of disappointment" at the weakness displayed by the government's opponents. To offset the bad impression Adams went ahead himself to attack the positions maintained by Evarts and Curtis, no doubt hoping that he might sway the court.

He charged that Curtis had not fairly faced the true issue in the case. "Our Constitution, with its doctrine of limited powers, was supposed to have settled the principle that our Government, unlike most other governments, could legally do only those acts which of right might be done; that there was in it, speaking in general terms, no supreme power, even in the last resort, to make wrong right or false true [or to say] that a piece of paper is equal in value to a pound of gold." As for Evarts' contention that "the safety of the state" was the paramount criterion "in times of national peril," the principle carried to "its logical conclusion" would allow Congress to abolish private property. "Mr. Evarts dashes at the obstacle of private rights, and, in overthrowing it, overthrows the whole fabric of our government."

Evarts bore the attack good-naturedly and his opinionated young friend continued to be a welcome guest at his house. Here, in December 1868, Adams met Moorfield Storey, then twenty-three, who had come to Washington as Sumner's private secretary. Storey's "Autobiography" describes the encounter with agreeable frankness: "Calling there one evening shortly after I reached Washington, I found a strange young man there who was monopolizing the conversation, as it seemed to me, and laying down the law with a certain assumption. I took quite a prejudice against him during this brief acquaintance, but the next day I met him in the street and he was so charming and his voice was so pleasant that my prejudice vanished, and we formed a friendship which was lasting." One of the fruits of that friendship seems to have been Henry's resumption of friendly relations with Charles Sumner; but as earlier with Henry's brother, Sumner avoided any interest in news of the ex-Minister. At occasional dinners at Sumner's home Henry gave full rein to his passion for argument, but Storey though a half-dozen years his junior was his match in violence of expression. "He would drop in and get into a fight with me," Storey recalled, "and we two would be so ill-mannered as to monopolize the whole conversation." One bone of contention was the moral condition of American politics. Henry now ten years removed from the idealisms of his Class Day Oration, took the gloomier view of the present, whereas Storey saw signs of improvement since the "intrigues of Jefferson and Monroe." A little exasperated by Adams's pretensions, Storey consoled himself with the thought that "it is very easy to make sweeping assertions, and to none more easy than the gentlemanly democrats of Boston, who have always gloried in their ignorance of politics." Curiously enough, Adams long afterwards remembered his combative friend as "a dangerous example of frivolity." [9]

The collaboration which Henry and his brother had so often planned now began in earnest and a constant stream of letters passed between Washington and Boston. Each week Henry also reported to his father on the state of affairs, carefully numbering the letters in his Date Book like diplomatic dispatches. As Charles kept "incessantly" after him for "documents, especially about railway matters and nice bits of cor-

ruption," he besought Representative Garfield to provide him
with "rubbish enough to supply his too active digestion." [10]
Charles had already allied himself with a number of the reform
groups, including the American Association of Social Science,
of which he was now treasurer, and was becoming widely
known in reform circles.

Their letters bubbled with all manner of comment and report
of work in progress. Henry did not care for the new administration, as it was shaping up, but he felt confident he could
"get on without it." Still his activity as a lobbyist and anonymous newspaper correspondent did not satisfy his sanguine
hopes. "I want to be advertised and the easiest way is to do
something obnoxious and to do it well. . . . I can work up
an article on 'rings' which, if *published in England,* would, I
think, create excitement and react through political feeling on
America in such a way as to cover me with odium. Wells says,
don't disgrace us abroad. I say, Rot! The truth is open to expression anywhere. No home publication will act on America
like foreign opinion." A few days' reflection on this project
opened up new vistas of sensationalism. "I can't do my 'Rings'
in short time. I am going to make it monumental, a piece of
history and a blow at democracy." At the same time he alluded
to another idea, "a popular article showing the practical expedients by which traders make a profit out of currency."
"These fields," he rejoiced, "are gloriously rich and stink like
hell."

While these projects were simmering, Atkinson, who was becoming the spokesman for the revenue and tariff reform group
in Boston, tried to get him to expose the international ramifications of the protective tariff. Adams countered: "I have already
a dozen ideas in my head which if elaborated would occupy
me years . . . To follow the protectionists over to England is
to go off on a false scent . . . The whole root of the evil is in
political corruption; theory has not really much to do with it."

What Atkinson thought of Adams's adherence to this notion we do not know. Apparently he never did succeed in interesting the young knight-errant in the new critical theories which were jumbling the neat pigeonholes of the old political economists.

Henry worked hard at his writing so that when he wrote to Reeve on February 1 he probably enclosed the completed manuscript of "American Finance, 1865-1869." The corrected proofs were lost in transit; but the article was much too good to hold and the editor ran it in the *Edinburgh Review* for April 1869, misprints and all. His enthusiastic opinion was passed on to Charles in Henry's best Chesterfieldian manner. "Reeve says it is the best article on American affairs ever printed in an English periodical and that he attaches the greatest importance to it. I don't see its astonishing merits, nor will you. But I hope it will make me unpopular. Q. E. D."

His panoramic view of American politics gave as grim a view of the subject as a Brady photograph of a littered battlefield. In his succeeding articles he would do little more than deepen a shadow here and sharpen a highlight there, and each would be developed in the Calvinistic brimstone of his rhetoric. Grant was about to take office. How should and could he run the government? Adams attempted to advise him and the new Congress. In every department of government reform was desperately needed. The reports of McCulloch and Wells showed that the Treasury was mired in difficulties. Secretary McCulloch, though capable, lacked the political stature and the force to carry out reform. As a result, the leaders of Congress deluded by "moral abstractions" embraced "the wildest financial theories," in defiance of the laws of economics.

Congress had betrayed its pledge to the Treasury by failing to order the resumption of specie payments. Pendleton of Ohio had built a formidable "greenback party." And Andrew Johnson had been naive enough to propose a scheme of converting bonds into annuities, in the belief "that creditors were influenced by ideas of abstract justice."

Bad as the currency system was, the tax system was even worse. "After three years of experience, the fact was no longer to be disguised that the whole revenue system was a mass of corruption, intolerable even in America where public opinion tolerates abuses such as would excite in England a revolution." In their analysis of the whiskey tax frauds Wells and Rollins had shown that corruption was not limited to Washington. Protected by public apathy the great corporations bought legislation at the state capitals as if at a sheriff's sale; privilege went to the highest bidder—or any bidder. At the bottom of this wholesale corruption were the regular party organizations. Reform to be effective must strike directly at them.

The tariff system, resting as it did on fraud and corruption, was a mass of inequities, the whole tendency of which, as Wells's statistics showed, was to hasten the process by which the rich were getting richer and the poor poorer. It was notoriously true that "No bill that *has money in it*" could become law until tribute was paid to the appropriate "ring" of lobbyists whose creatures in Congress daily repaid their election debts. For a price, "every petty manufacture" could "get its rag of protection" to clothe its avarice. It was obvious, Adams concluded, that "the great responsibility of the new administration is to itself and not to the world. The best Americans are looking to it with the deepest anxiety to save the country as far as possible from its dangers by effecting a reform the principles of which we have pointed out; but if the hope is disappointed, even though the country should go on increasing its wealth and power more rapidly than ever, the world will have the right to believe that neither the skill of government nor the virtue of American institutions has any share in the result, except so far as the nation is receiving and exhausting advantages left to it by a past and purer generation."

The article may not have attained the "monumental" quality which he had planned for his elaborate study of the rings, but it was a hard enough "blow at democracy" and should have brought him the "public horror and disgust" in America for which he had aimed. Contrary to his expectations the article seemed to infuriate no one in America, and in England Gaskell ran across only one review of it, in a Yorkshire paper. Adams saw the face of his chiefest enemy: apathy; but he was not yet disheartened.

Some of the winter's notes overflowed into another article written shortly after Grant's inauguration. He rushed it off to Professor Gurney, now chief editor of the *North American*, after adding the final touches in a burst of industriousness for "ten hours a day for four days, and politics on top of it." Thinking it might embarrass his father, he asked Charles to "stand by to whitewash him of all responsibility." The article, published in April, frankly imitated Lord Robert Cecil's "annual review of politics" in the *London Quarterly*. Its opening paragraphs dramatically pictured the Congressional gauntlet through which all legislation had to run.

Within the walls of two rooms are forced together in close contact the jealousies of thirty-five millions of people,—jealousies between individuals, between cliques, between industries, between parties, between branches of the Government, between sections of the country, between the nation and its neighbors. As years pass on, the noise and confusion, the vehemence of this scramble for power or for plunder, the shouting of the reckless adventurers, of wearied partisans, and of red-hot zealots in new issues,—the boiling and bubbling of this witches' caldron, into which we have thrown eye of newt and toe of frog and all the venomous ingredients of corruption, and from which is expected to issue the future and more perfect republic,—in short the conflict and riot of interests, grow more and more overwhelming; the power of obstructionists grows more

*and more decisive in the same proportion as the business to be
done increases in volume; the effort required to accomplish
necessary legislation becomes more and more serious; the machine groans and labors under the burden, and its action becomes spasmodic and inefficient.*

For all its rhetoric the description was hardly overdrawn. In
the privacy of his diary Secretary of the Navy Welles recalled
"the decadence of the Republic of Rome and the degeneracy
of her people." Like Adams he thought "the press is terribly at
fault."

The most pressing matter of "public anxiety" was the need
for restoring to the president the privileges and functions which
had been usurped by the Senate in its struggle with President
Johnson. If public opinion acted upon the lower house with
sufficient strength, the Senate could be reduced to its proper
place within the political order. For the rest, Adams reiterated
the lessons of the earlier article. Reform must come quickly.
"Perhaps not more than one member in ten of the late Congress ever accepted money," yet this criminal element held the
balance of power and the spoils system fostered this criminal-
ity. Representative Jenckes's bill for the reform of the civil service was the very least to be aimed at. But these were all details;
whereas the lesson to be drawn from the past ten years of
"incessant confessions" of ignorance, impotence, and even im-
becility was that the administration must adopt a coherent
scheme of reform for the entire government.[11]

The *Nation* thought that Adams's "Session" deserved "to be
styled statesmanlike," and even more enthusiastic was the opinion of the Springfield *Republican*. In an editorial entitled "Another Adams" his friend Sam Bowles wrote:

*Among the officers of the new "Reform League" at Boston
may be found the names of three Adamses—all sons of the late
Minister to England, and great-grandsons of the second presi-*

*dent of the United States. Two of these names are well known
to the people of the country, both in the present and the
past generations,—John Quincy Adams and Charles Francis
Adams. But the third—Henry Brooks Adams—designates a
young gentleman who has quite as good a chance of becoming
prominent in the future politics of the country as either of his
brothers, although he is yet but little known . . . He had the
reputation of being one of the three best dancers in the capital . . . The fruit of his winter's studies in Washington now
appears in the April* North American—*a long and brilliant paper on "The Session," in which, with some conceit and some
pedantry, but with more ability than either, he reviews the
doings and omissions of the last session of the Fortieth Congress.*[12]

The delicious taste of fame, fame earned by his own effort,
gave him fresh confidence. Forgotten was his recent dread of
newspaper personalities in the realization that what the press
took away it could also restore in lavish measure. Enclosing the
editorial in a letter to Gaskell, he exulted, "For once I have
smashed things generally and really exercised a distinct influence on public opinion by acting on the limited number
of cultivated minds . . . But you see I am posed as a sort of
American Pelham or Vivian Gray. This amused me, for you and
I have always had a foolish weakness for combining social and
literary success." The sequel, as he said, was a sufficiently curious illustration of journalism. "This leader was condensed into
a single paragraph . . . and copied among the items of the
column 'personal' all over the country. In this form it came back
to New York . . . Now however the paragraph is compressed
to two lines. 'H. B. A. is the author of articles etc., etc., etc. He
is one of the three best dancers in W.' . . . I am in an agony
of terror for fear of seeing myself posted bluntly: 'H. B. A. is
the best dancer in W.' This would be fame with a vengeance." [13]

Encouraged by the stir he had caused, he thought he might duplicate in America Lord Cecil's success in England. "If the future goes straight," he promised Charles, "I will make my annual 'Session' an institution and a power in the land."

Fundamental Principles

Advertisement could not confer power overnight, however, and his influence in the Cabinet rapidly deteriorated after the inauguration of Grant. The new Cabinet disappointed and bewildered the men of good will who had counted on strong and progressive figures. Brilliant David A. Wells was passed over in favor of unimaginative George S. Boutwell for the Treasury. Adams could still count on sympathy from Jacob D. Cox in the Interior and Attorney-General Ebenezer Rockwood Hoar, but these friends of reform had only a precarious hold on Grant's confidence. By March 11 his disillusionment was complete. "It's the old game with fresh cards." To John Hay, then visiting Washington, he said: "A man asked me the other day if I had been at the White House lately, and I told him, No. I want to remember that house as Lincoln left it." [1]

Washington became almost pleasantly deserted when the Fortieth Congress adjourned, and he lingered on until the end of June much recovered after the recurring fits of wretchedness imposed by his "disordered liver," but spring found him "thin and bearded and very—very bald." Reeve of the *Edinburgh Review* called on him for another article on the "changes of the American Constitution which have resulted from the war," but he begged off because, he believed "The essential and fatal changes in our Constitution were not the result of the war, but of deeper social causes, which each need a volume to discuss." He turned instead to a more literary project, a biography "which may grow to be three volumes if I have patience to toil" based on materials left by "an ancient lady of our house."

The project—presumably a life of his grandmother Louisa Catherine Adams—languished and died long before the summer was out.

Such loitering "Bohemianism" displeased the more dynamic Charles and he read Henry his customary lecture. Sick of being pushed, Henry fought back: "I see you are getting back to your old dispute with me on the purpose of life . . . I will not go down into the rough-and-tumble, nor mix with the crowd; nor write anonymously, except for mere literary practice . . . You like the strife of the world. I detest it and despise it. You work for power. I work for my own satisfaction. You like roughness and strength; I like taste and dexterity. For God's sake, let us go our ways and not try to be like each other." His words came almost like a cry from the heart of the young undergraduate who had longed for a quiet literary life and now found himself harnessed to the family cart without the will to break the traces.

Henry knew that he had committed his forces. Yet he had no clear idea of what he really wanted; "certainly not office, for except very high office I would take none. What then? I wish someone would tell me?" The question being insoluble, he turned more practically to plans for the next session of Congress.

As Wells and Garfield were still in town he invited them up to Quincy for a council of war, sending word on to Charles, "We will have Atkinson too, and Greenough, and cut out our work." Garfield, pleased to make the pilgrimage to Quincy, sent home a glowing account of his hosts, among whom he described Henry, "rapidly rising as a clear and powerful thinker and writer." [2] Though Henry belittled these Quincy conferences as "humbug," they nonetheless fixed the direction of his work. Toy as he might with literary biography or Tyrone Ghost Stories, he rose to more sober business, burying himself under "acres of books" on finance. He also proceeded to salvage materials left over from his two April "firecrackers." At the end of August "after three months hard labor" he was "accouché of another ponderous article . . . bitter and abusive of the Administration," which upon publication in the October 1869 *North American* he planned to distribute as a pamphlet to the members of Congress; its title was "Civil Service Reform."

In essence, the article was an extension of remarks to the April "Session," but now Adams attempted a more philosophical analysis of his subject. "Laying aside the usual arguments in favor of civil service reform,—arguments drawn from finance or from administrative convenience,—this essay will attempt to show that the soundness and vigor, nay, even the purpose of the reform movement, must depend upon its recurrence to the fundamental principles of the Constitution." At the heart of the Federal system was the principle of the due separation of powers. Grant's troubles stemmed from the neglect of that principle. Now that the Republicans had "shut the door to reform" by surrendering to the Senatorial system of appointments, it would be necessary for the people "to act outside of all party organizations." With "a sympathetic public behind him" ready to discipline the Senate he could himself reform the civil service. To create that sympathy "there is no way but to attack corruption in all its holes, to drag it before the public eye, to dissect it and hold its diseased members up to popular disgust, to give the nation's conscience no rest nor peace until mere vehemence of passion overcomes the sluggish self-complacency of the public mind."

William S. Robinson, a friend of Senator Sumner, hastened to the defense of the Senate in a "Warrington" letter to the Springfield *Republican*. The spoils system, he blandly asserted, occasioned "no such distress as justifies the rhetoric which Mr. Curtis and Mr. Henry Brooks Adams have been wreaking on the subject . . . We have had nothing since Burke's description of the ravaging of the Carnatic by Hyder Ali, which equals

Mr. Adams's description of the scenes at Washington after Grant 'gave way' as he styles it." It was a shrewd blow and Henry Adams was all too vulnerable. His admiration for Burke's eloquence had trapped his pen as "Warrington" was quick to show in the following parallel: Edmund Burke— "Then ensued a scene of woe, the like of which no eye has seen, no heart conceived, and which no tongue can adequately paint"; Henry Adams—"Then began those cruel scenes which for months reduced the city of Washington to such a condition as is caused by an ordinary pestilence or famine." The *Nation,* on the other hand, declared approvingly that Adams had "by this and one or two similar essays made for himself an enviable reputation as a courageous politician in the best sense of the term, and as an excellently clear and forcible writer. Indeed he may be said to be at times a little too forcible." [3]

Unfortunately for Adams's expectations the pamphlet expired in a web of silence. No allusion to its prophetic thunders occurred in the course of the long debates on the civil service bills. It did catch the attention, however, of Secretary Cox, as Adams learned from Garfield. Since he had made no mention of Cox in the course of "Civil Service Reform," although Cox was deeply interested in the subject, he felt he ought to justify the omission. "The fact is," he wrote Cox, "you are the reserved force, the silent agency by which I hope this contest is to be decided." The time had now come for Secretary Cox to act: "Give the country a lead!" Adams pleaded. "We are wallowing in the mire for want of a leader. If the administration will only frame a sound policy of reform, we shall all gravitate towards it like iron-filings to a magnet." [4]

Henry stayed on in Quincy until about the end of October "reading Gibbon and wasting time" while waiting for the Forty-First Congress to convene. In his leisure he sent Gaskell a "disquisition" on American literature for use in a lecture. Although he thought it "a pity you can't quote some choice lines from Walt Whitman," his recommendations move rather conventionally through Bryant, Longfellow, Lowell, Hawthorne, and similar writers. "There is nothing very new," he concluded. "We have no writers now." Passed over silently were men like Emerson, Thoreau, Holmes of the *Autocrat,* Melville, Poe, Lanier, and the new voices that were speaking out in the *Atlantic,* the *Galaxy,* and the *Overland Monthly:* Bret Harte, Mark Twain, William Dean Howells, and Henry James. In literature as in economics and politics, Adams's tastes were solidly conservative.[5]

Amid the ancestral surroundings in Quincy he grew more and more restless for he found life to be "on too small a scale for man to sustain it without idiocy." Even though he wrote that "my two brothers and I are up to the ears in politics and public affairs," he still felt the want of "excitement." For the third time his eldest brother was running for the governorship on the Democratic ticket; but the preceding two defeats had perhaps engendered a certain degree of stoicism in the family.

Relief was in sight, however, for on Black Friday, September 24, 1869, an attempted corner in gold bullion crashed spectacularly about the ears of certain gentlemen of the Erie Railroad, opening up for the Adams brothers the exploitation of a new vein of knavery. Charles had already published two articles on the intricate depredations that made up the current history of the Erie and could direct his brother to the most savory sources. Henry proceeded to New York and there dined at Samuel Barlow's with William Evarts, already ex-attorney-general of the United States. Barlow, who was also counsel for McHenry, one of the more imaginative railroad financiers, "told some instructive stories about Erie, Atlantic and Great Western, etc." Equally instructive was his remark that Jay Gould and Judge Barnard "had expressed the intention of taking hold" of Charles if he ever came to New York. Charles had in fact been in New York at that time to address the Social Science Association on election frauds, but he escaped the notice of his angry victims. In New York Henry saw "many editors; some thieves; and many more fools," one of the "thieves" being James Fisk himself.

He arrived in Washington on November 1 eager to put his hands again on the levers of power, the lurid scenes of "The Gold Conspiracy" already shaping up in his mind. He sketched out the drama in a letter to Reeve of the *Edinburgh Review*, expecting an instant acceptance. Nothing happened. Irritated by several weeks of silence he damned him as "that animal Reeve" and asked Palgrave to take the article to the *Quarterly*. He promised Gaskell, "I will make my article SUPERB to disgust Reeve." On December 13, shortly after Congress met, Garfield as chairman of the House Committee on Currency and Banking, was directed to investigate the New York gold panic of September 24. With an official investigation under way, Henry would now have to possess his soul in patience until the report should be ready, but he could console himself that no other writer was more strategically placed.

There was much to do in Washington. He now threw himself with intense energy into his journalistic work, contributing "about two articles a month in the *Nation*," and sedulously cultivating his sources of information. His friend General Badeau, with whom he maintained a joint establishment for a time, presented him at the White House where he found the President and Mrs. Grant so ill-at-ease that "it was I who showed them how they ought to behave." To economize his energies he never dined with anyone unless there was "something to be gained" and soon felt that he was wound up "in a coil of political intrigue and getting the reputation of a regular conspirator."

Few of his contributions to the *Nation* can be positively identified but there is no longer reason for regarding his assertion with suspicion.[6] The first of the series was probably "Men and Things in Washington," dated from the Capital, November 20,

1869. Assuming the guise of a foreign diplomat in Washington, Adams made a devastating survey of Washington society. "There is not a theatre, except by courtesy; there is no art, no music, no park, no drive, no club. Good dinners are rare indeed, and good wines are, for the most part, held firmly in diplomatic hands . . . The worst annoyance of all is that society itself exists only in disjointed fragments, and that there is no established centre of intelligence and social activity. If a diplomat wishes to mix with congressmen and senators, or with the persons of note who are incessantly passing through Washington, it is ten to one but that he must hunt them to their separate retreats, and after cornering one in a lobby, or running another to earth in his office, or his bed-chamber, or at his favorite barroom, each and every one stands at bay when approached, as though he had something to lose in this rare contact with a gentleman, and nothing, at his time of life, to learn or to hope from cultivated society." His concluding suggestion that the administration must assert "large principles" and accept "high responsibilities" was almost lost in an elegant tangle of Gibbonesque clauses. The whole was well-seasoned with literary allusions: an apt jest from Disraeli's *Tancred*, and passing mention of the Greek chorus, the heroes of Corneille, and the stage customs of Shakespeare's day.

His highly individual style likewise colors "A Peep into Cabinet Windows," dated December 12. Its opening paragraph gives an early version of one of his favorite allusions. "The liberal reformer's idea of an administration after his own heart would no doubt at the outset assume, like Voltaire's satire and Presidents' messages, that in this best of all possible worlds this nation is the best of possible nations, and by consequence this Administration the best of all administrations, past or to come." Again the satire of the Administration was squeezed through a labyrinth of subtleties. Grant's message with its contradictory recommendations came in for some hard knocks, but as usual with Henry Adams, Secretary of the Treasury George Bout-well suffered most. "The wide space his figure should fill is left void and blank, except for the vague outline of half-a-dozen blurred and unrecognizable forms which seem at one time to resemble the President, at another some one secretary, or, again, some face not known in Washington. Is it possible that Mr. Boutwell has of his own accord obliterated himself in this shocking manner, or has he been compelled to do so? Or, most alarming supposition of all, is this droll figure of four and a half per cent, all that really exists of Mr. Boutwell anywhere in this world; his whole and identical image; what Carlyle would call Mr. Boutwell's mysterious ME, standing in the conflux of eternities?"

The rejection by the Senate of his friend Hoar's nomination to the Supreme Court called forth the mordant contempt of his next dispatch, "The Senate and the Executive," in which he lashed at the "cabal" made up of the "dregs of the Senate Chamber" who struck "not so much at the possible judge as at the actual Attorney-General." Hoar's guilt had been that of appointing "competent men to office regardless of political combinations." Unless Grant directly met the challenge to responsible government, the dangerous notion of "ministerial responsibility to the Senate," would become firmly entrenched. This theme ran through several more articles which, though not dated from Washington, seem also to be his. Thus on January 27 "A Political Nuisance" pounded Sumner and the arrogant senatorial school of Reconstruction. "Mr. Sumner travesties himself, and turns the noble common-school system of New England into a caricature of his old appeals to the moral force of liberty." Since one of the allusions to the "nagging of the Executive" led to "some unfavorable comment," an ironical article soon followed: "Mr. Dawes—President Grant—General Butler."

Adams's journalistic activities extended also to William Cullen Bryant's New York *Evening Post* and to at least one other New York paper, possibly Manton Marble's *World*, which was especially friendly since John Quincy's conversion to the Democratic party. As he dodged about with pen in hand he would lampoon "a highly respectable gentleman who is a friend of mine" for one paper and then "write another long article abusing everybody for another paper." Only one of these gadfly attacks provoked a visible response. Dennis McCarthy of New York, a "high tariff" member of the Ways and Means Committee brought up on the floor of the House Adams's unsigned editorial in the *Post* accusing him of corrupt behavior. McCarthy blustered that "the whole statement of facts is untrue," but especially objected to Adams's accusation that "one of these corporations, grown wealthy and powerful, sent a gentleman to Congress with the understood purpose of representing its interests. Its influence caused Mr. McCarthy, one of its chief shareholders, to be elected . . . He has done his work with success. His single vote has saved salt from interference in committee. Nor is this all. Besides protecting his own especial preserve he has found time to watch tenderly over every other monopoly and to lend his active and unusually decisive aid wherever the interests of a privileged class were threatened. His career in Congress has been eminently useful to his friends." In an irate speech that occupied a column and a half of the *Globe,* McCarthy rejected the back-handed compliment because it was "said in a hypocritical manner and with malicious intent." The editor of the *Post* continued the exposés.[7]

Stimulated by the "incessant excitement" and "the rough-and-tumble Bohemianism" of his position, Adams felt himself "tending more and more toward journalism," from which he got "a little money to buy gloves with, and a certain power to make myself felt." As his colleagues in Wells's circle preëmpted his services in behalf of revenue reform he had little time left for more extended literary efforts. For instance, after the publication of the 1869 Report on the Revenue, Francis A. Walker exhorted his chief, Wells, "Now it must be followed up. . . . *You* must write, Henry and Charles Adams must write, Atkinson must keep pegging away." [8]

In spite of this pressure upon him Adams managed to write one long article for his standby, the *North American,* taking the assignment to relieve his friend Walker. In 1869 when Spaulding brought out his *History of the Legal Tender Paper Money Issued during the Great Rebellion,* Adams, acting for Editor Gurney, got Walker's promise to upset Spaulding's thesis that the Act had been necessary. Before Walker could finish his study, the Supreme Court, in a four to three decision of *Hepburn* vs. *Griswold* declared the Legal Tender Act unconstitutional so far as it "applied to contracts made before its passage." By chance, Walker was appointed Superintendent of the Ninth Census the very day the case was decided, February 7, 1870. On the same day Grant significantly added two justices to the court, appointments which later led to the charge of "packing." At this point Walker turned over his notes to Adams. The ruling of the court necessarily changed the bearing of the article; whereas before it might have persuaded the court to invalidate the Act, now that that point was won, its chief value would be to support the hand of the meager majority and thus protect the court from the threat against its independence. How real that threat was became evident in March, when on the rehearing of the case, Attorney-General Hoar began his thankless job of browbeating Chief Justice Chase who had spoken for the majority.[9] To protect the new Census Superintendent from reprisal, Adams signed himself as the sole author of "The Legal Tender Act" when it came out in the April *North American.*

In it he poured out upon Congress, the late Thaddeus Stevens, and the ill-starred Spaulding, what he had carefully designed as "a piece of intolerably impudent political abuse." The article may have profited from the hand of Walker, but its voice was unmistakable Adams. For a classical parallel, he exhumed Cleon, the egregious leather-seller statesman, and then he launched his Aristophanic attack upon the recently dead Stevens. "To say that Mr. Stevens was as little suited to direct the economical policy of the country, as a naked Indian from the Plains to plan the architecture of St. Peter's or to direct the construction of the Capitol, would express in no extreme language the degree of his unfitness."

Having thus paid his respects to the dead, he turned to the living: "The intellect of a Congressman, gifted with no more than the ordinary abilities of his class, is scarcely an interesting or instructive subject of study . . . But when an accidental representative, though he may possess neither breadth nor force, is able to carry . . . a measure . . . as Mr. Spaulding claims to have done . . . the character of that person's mind and the facts of his life cease to be matters of absolute insignificance." Spaulding, who had acquired his financial education "in shaving notes at a country bank," had rejected as discreditable "the simple proposal" made by the New York bankers "that the government should sell its bonds in the open market for what they would bring." This "response," said Adams, "contains the most extraordinary revelation of ignorance and folly that has been offered in a century past by a principal legislator for a great nation." By taking pride in the Legal Tender Act, Spaulding placed himself "beyond the reach of criticism," for he "perverted legislation into an instrument of fraud, and is proud of it." Yet Spaulding, like Stevens, would have to face a higher judgment than that of his contemporaries, "the judgment which historical criticism must inevitably exact for the betrayal of principles."

Once again, in the course of the essay, Adams affirmed his faith in the dogmas of laissez-faire economics. It was indifferent, he held, whether the power to issue legal tender "has or has not been conferred by the Constitution upon Congress. Its financial merits are not a subject for lawyers or even for judges. These happily rest on principles deeper than statute or than constitutional law." For a devotee of the Constitution he was uttering more dangerous doctrine than he knew, as the argument of the higher law had already cost a civil war and could more properly be asserted by non-idolators. The opinion recalls his 1861 comment that it was futile for the stone masons' union to strike for a nine-hour day as such matters are "regulated by rules which are beyond the just range of mere enacted law." Obviously the grand idealism of his Class Day Oration had undergone a significant change. There were now incorporated among "those great truths, to the advancement of which Omnipotence itself has not refused its aid," the principles of classical political economy.

Spaulding tried to vindicate himself by circulating a letter to the press denying that he had claimed credit for forcing the Legal Tender Act through Congress. Adams riposted in a long contemptuous letter to the *Nation,* citing Spaulding's self-laudatory *Financial History of the War* against its author. "What the *North American Review* meant to protest against was his notion that a public man may be permitted to receive credit for the authorship of a dishonest measure, the ultimate results of which he had neither the knowledge nor the capacity to foresee, and the effects of which he joins in deploring." A farcical aspect of the controversy appears in a letter to Gaskell. "We call each other fool and idiot in the papers, and carry on a very friendly private correspondence meanwhile."

The editor of the *Nation* enjoyed the exchange of incivilities, for, he said, Henry's article recalled to him Rufus Choate's allusion to the " 'peculiar powers as an assailant . . . an instinct for the jugular and carotid artery as unerring as that of any carnivorous animal,' " which had belonged to old John Quincy Adams. "Mr. Henry Brooks Adams justifies his title to the family name . . . Mr. Spaulding, having imprudently shown signs of life before Mr. Adams left the field, has been despatched by having several large stones thrown on him." A subsequent contributor observed sympathetically that Spaulding "is certainly no match in single combat for Mr. Henry Brooks Adams." [11]

Fame and Its Aftermath

Garfield brought out his report on the gold panic on March 1, 1870, but most of the findings were already known to Henry Adams who had been active "pulling wires." Garfield had naturally turned to the Adams brothers for advice on how to unwind the skein of knavery which Gould and Fisk had spun. Too busy to accompany Garfield to New York, Henry had sent him general encouragement: "My idea is that there are only three or four points in the whole story which need new light, and your preliminary object must be to satisfy your own mind what they are, and when found, to disregard everything else." As more concrete help he enclosed a letter of introduction to a strategically placed member of the Stock Exchange.[1] With the mass of testimony before him, he was now ready to begin on "The New York Gold Conspiracy." His annoyance with Editor Reeve was not dissipated by Reeve's belated excuse that he wanted "nothing controversial about currency." Knowing that the article would contain "a good deal of libellous language" which he could not afford to publish in the United States, Adams persisted in seeking an English publication. If the London *Quarterly* would not have it he intended to take it to England himself "to find some editor" who would. The mission would fit in to his long-deferred plan for an English holiday.

As usual the work was incessantly interrupted by his activities as a lobbyist. Early in April, after having helped organize "our forces" as "lieutenant" for Wells, he was pitched into "the middle of a battle over revenue reform, free trade, and what not." In pursuit of the plan to form a new party on these issues, he entertained about a dozen of the leaders in his rooms. They effected "a close alliance" and agreed to "hold a secret but weighty political caucus on the 18th to which our friends—the small number of high panjandrums—from all quarters, are to come." Profiting from Lord Houghton's example, Adams played host to the caucus at a "breakfast" which was attended by "half of the greatest newspaper editors in the country." As he dramatically reported to Gaskell, "The world was staked out to each of us and the fulness thereof, and the foundations of Hell were shaken." How he worked his claim is suggested by a letter of May 12 to Congressman Garfield: "Will you see Schurz and urge him to introduce your Civil Service Resolution into the Senate as a joint Resolution? Don't let him know that I suggested it, as he might not like my interference. I had a little talk with him about it yesterday and he seemed not indisposed." [2]

Completed toward the end of April, the "Gold Conspiracy" was packed off to Francis Palgrave for the *Quarterly Review*. It had weathered not only his work as a "conspirator" but also his recurring attacks of irresolution for he was "still at a loss to know what the devil I want, or can possibly get, that would be an object, in case my friends come into power." The strain of his position was beginning to tell. James S. Pike, encountering him one night at Sumner's, thought he had "grown old since I knew him in London" and "crochety and conceited as well." [3]

Adams had threatened to make the gold conspiracy article "superb" and succeeded, but the success was again more literary than political. Try as he might, the reformer lost himself in the historian as he explored the "curious and melodramatic" aspects of the story. "It is worth while," he explained, "for the public to see how dramatic and artistically admirable a conspiracy in real life may be, when slowly elaborated from the subtle mind of a clever intriguer, and carried into execution by a band of unshrinking scoundrels." In a companion piece, "A

Chapter of Erie," Charles Francis Adams, Jr., showed how the great capitalists were driven to plunder each other as the area of speculation narrowed in the postwar world, until buccaneers like Drew and Vanderbilt became "two cormorants reduced to tearing each other." In his essay Henry provided a more spectacular illustration of this financial cannibalism.

Adams was quite right in suspecting libel. Nowhere before had his characterizations been so outspokenly venomous as those which introduced the two protagonists of his melodrama. The president and treasurer of the Erie Railroad, "Mr. Gould . . . was a broker . . . perhaps the very last profession suitable for a railway manager . . . There was a reminiscence of the spider in his nature. He spun huge webs, in corners and in the dark, which were seldom strong enough to resist a serious strain at the critical moment . . . It is scarcely necessary to say that he had not a conception of a moral principle . . . without entering upon technical questions of roguery, it is enough to say that he was an uncommonly fine and unscrupulous intriguer, skilled in all the processes of stock gambling, and passably indifferent to the praise and censure of society." The vice-president, "James Fisk, Jr., was still more original in character. He was not yet forty years of age, and had the instincts of fourteen . . . Personally Mr. Fisk was coarse, noisy, boastful, ignorant; the type of young butcher in appearance and mind . . . In respect to honesty as between Gould and Fisk, the latter was, perhaps, if possible less deserving of trust than the former." He established his offices adjoining the opera house, "and as the opera itself supplied Mr. Fisk's mind with amusement, so the opera troupe supplied him with a permanent harem. Whatever Mr. Fisk did was done on an extraordinary scale." Neither Balzac nor Alexander Dumas had "ever ventured to conceive a plot so enormous, or a catastrophe so original."

The development of the plot had the intrigue, passion, and

clamor of an Italian opera. Briefly the conspiracy which Adams so skillfully unravelled took the following course. Gould conceived the Napoleonic scheme of obtaining a monopoly of the floating gold supply so as to extort a fabulous price from persons who needed gold to meet contracts requiring payment in specie. Gould, Fisk, and their agents stealthily began to purchase gold. Anticipating that the Treasury might frustrate their scheme, the conspirators enlisted President Grant's venal brother-in-law, Corbin, in the expectation that Grant would be persuaded not to intervene. By the morning of Black Friday, gold had risen 25 points to 150. At this point Fisk stepped in to administer the *coup de grace*. He flooded the market with buying orders. Grant, meanwhile having been stirred to belated action, ordered Secretary Boutwell to sell $4,000,000 in gold. Amid incredible scenes of frenzy the price plunged from 165 to 138. Gould, forewarned of the Treasury's action, had secretly sold his holdings during the final upsurge of the market, at the very moment when his associate Fisk was frantically buying everything offered. The investigating committee, accepting Fisk's cries of injured innocence as genuine, concluded that Gould had "determined to betray his own associates," and Adams accepted this interpretation. It was passing strange however that the bankruptcy of the Erie's brokers did not touch Fisk for he had foresightedly induced them to buy at their own risk.[4]

As Adams knew that it was Garfield's private opinion that the trail of scandal "led into the parlor of the President," his narrative carried a strong imputation against Grant's probity.[5] Once the author should be discovered he must thereafter count on Grant's personal hostility. He laid up other hatreds among the gentlemen of the bar. Only the steady demoralization of the legal profession had made it possible for the Erie Ring to obtain the services of Mr. David Dudley Field, "a gentleman of European reputation." This paragon of American lawyers had rendered services to the Ring which could "not be measured even by the enormous fees their generosity paid him." According to the "gossip of Wall Street" Field had "a silken halter round the neck of Judge Barnard, and a hempen one round that of Cardozo." Adams also charged that the article which Gould had induced Grant's brother-in-law to write on "the financial policy of the Government" had been "inserted in the New York *Times* through the kind offices of Mr. James McHenry," who was Samuel Barlow's eminent client. Both Mr. Field and Mr. McHenry were deeply offended; the former, as we shall see, took his grievance to the press; the latter threatened to sue for libel.

Adams's conclusion paralleled his brother's that the financial manipulators had introduced "Caesarism into corporate life." [6] Foreseeing the age of trusts and cartels, he declared, "The belief is common in America that the day is at hand when corporations far greater than the Erie—swaying power such as has never in the world's history been trusted in the hands of mere private citizens, controlled by single men like Vanderbilt, or by combinations of men like Fisk, Gould, and Lane, after having created a system of quiet but irresistible corruption—will ultimately succeed in directing government itself."

With the sulphurous manuscript safely on its way to England, although only a fortnight remained before Adams's own departure for England, he now took up his self-imposed chore of writing the epitaph of the first session of the Forty-First Congress for the July *North American*. The "Session" of 1870, like that of 1869, assailed the shortcomings of the political "system." It recurred to the somber thesis of de Tocqueville, already a commonplace of the critics of democracy, that the American system might not survive its crucial test, when the test finally came. The war had abrogated the Constitution, but with Grant the opportunity had come to restore to working order the intricate system of checks and balances. Grant had disappointed the expectations of the people because his "personal notions of civil government were crude" and "his ideas of political economy were those of a feudal monarch a thousand years ago."

Following the pattern of the earlier "Session" he ran through the various issues which had strained the institutions of government. The Reconstruction measures showed that the reserved powers of the States were hereafter to be exercised "only on good behavior, and at the sufferance of Congress." The proponents of revenue reform had met with "cold indifference" and "hostility" from the Administration. Tariff and currency reform fared little better. But most disastrous in his eyes was the attempt to pack the court after the February 7 decision in *Hepburn* vs. *Griswold*.

In foreign affairs the disgraceful contest between Grant and the Senate over the St. Thomas treaty, San Domingo, and Cuban intervention, a contest in which Sumner had "issued his orders with almost the authority of a Roman triumvir," [7] showed a policy "so unsteady" as not "to command respect." What was to come of this dangerous tug of war between the executive and the "omnipotent senatorial authority," he proposed to leave "for future annual Reviews to discuss."

What was even more disheartening, however, was the fact that there was rising, through the agency of the railways and the telegraph, such a concentration of "social and economical forces" as to render the governmental system itself obsolete. The work of political reformers must be lost labor, for "while one monopoly is attacked two are created." In the face of this deluge of evil, he retreated to higher ground: "Nor is it here intended to point out, or even to suggest, the principles of reform. The discussion of so large a subject is matter for a lifetime, and will occupy generations. The American statesman or philosopher who would enter upon this great debate must make his appeal, not to the public opinion of a day or of a nation, however large or intelligent, but to the minds of a few persons who . . . attach their chief interest to the working out of the great problems of human society under all their varied conditions."

The mood was that of disenchantment; and the reverberations from Burke and Gibbon and Carlyle rang through the carefully wrought sentences with a sublime clangor. The *Nation* called it "the most widely attractive article in the *North American Review*," though the tone of some of it was "too unmitigatedly and severely fault-finding and critical." Several newspapers reprinted the article in its entirety and it was rumored that the Democratic National Committee planned to issue it as a pamphlet and circulate 250,000 copies. He could now safely assert that he had a "tolerably large audience," but he should have added that many members of it were very angry Republicans. One of these, Senator Timothy O. Howe of Wisconsin as spokesman for the party press, published an elaborate reply in the *Wisconsin State Journal*.[8]

Senator Howe considered Adams's effort the first broadside against the Republican party in the campaign of 1870, and with reason, for the article had been released to the Democratic party press in advance of publication. Matching gibe for gibe Howe sneered, "The author is proclaimed to be not only a statesman himself but to belong to a family in which statesmanship is preserved by propagation—something as color in the leaf of the Begonia, perpetuating resemblance through perpetual change." The senator's objections to the matter of the article were more colorable, for it was plain that Adams had rushed into a number of inconsistencies in the heat of his denunciation. These Howe exposed point by point, with such senatorial embellishment as the following: "Nowhere is the charge of incontinence so freely fabricated or so fiercely hurled as in a brothel." He also paid a left-handed compliment to Adams's portraiture. "But he demonstrates his ability to sketch character as Hogarth did, by the facility with which he produces caricature . . . The author has a talent for description and a genius for invention. He might succeed as a novelist. He must fail as a historian." Despite its scurrilous character, Howe's reply had point and force, but Adams chose not to make a rejoinder.

He was more than ready for his English holiday. Retrospect had long added its charm to England. Amidst the social short-comings of Washington and Quincy he yearned for his English friends and the amenities of British life. Six months after he had left London he had begun to plan his return visit: he thought he would "pretty certainly come over in June, 1870." Mired in the study of corruption in high places, he was eager to "go to England for a moral bath." When the time came to book passage his pleasure at the thought of rejoining Gaskell, his "Caro Carlazzonuccio," swept sardonic reflections out of the way. With touching directness, he wrote to his most intimate English friend, "I come over largely to see you and yours."

He reached England about the first of June for his rendez-vous with Gaskell, but in a few weeks his vacation was suddenly cut short by news of his sister Louisa's dangerous illness at Bagni di Lucca. He hurried off to Italy, reaching his sister's bedside on July 1. At first he felt himself "a wretched coward" when he saw the symptoms of acute tetanus, but Louisa's extraordinary courage and good humor soon communicated itself to him and he was able to describe to Gaskell with gruesome vividness the mixed horror and gaiety of the sickroom. On July 13 however he wrote, "It is all over. My poor sister died this morning." The tragedy spoiled his appetite for any kind of merriment and he made his way back to England in a subdued mood.

As he passed through Paris late in July a week after the declaration of war against Prussia, he bitterly reflected that the war put "an end everywhere to any chance of carrying out a regular course of politics." Disgusted by Europe's imperviousness to "what we call progress," he wished that both nations could be simultaneously beaten. In England he hunted up some of "the political people still in London" for talk on politics and to determine the fate of his "Gold Conspiracy." The article proved surprisingly difficult to place. Finally "as a last experiment" he sent it to the *Westminster Review* just before his departure for America on September 3. Not until its publication in October was he aware that it had been accepted.

The summer was to be memorable for an even more significant train of events. Henry Adams had had a slight hint of what was afoot in the preceding April when Professor Gurney, recently elevated to the post of Dean of the Faculty at Harvard, offered him the editorship of the *North American*. Not wishing to exile himself from Washington he had declined the offer, though he stood ready to act as "official editor for politics." A few days after his arrival at Bagni di Lucca he received two letters which had pursued him across Europe. They were from Charles W. Eliot, the new president of Harvard College, and they offered him a post as assistant professor of history. With no family council to urge prudence upon him, he replied: "The offer you make me is not only flattering but brilliant, yet I cannot accept it. Two years ago I should have hesitated long before deciding, but having now chosen a career, I am determined to go on in it as far as it will lead me. Perhaps ten or fifteen years hence, if I am satisfied that my experiment has failed, I may be glad to make myself useful at Harvard if I am still wanted but for the present I can only thank you for the kindness of your proposal." [9]

Late in August, Gurney, still hoping that Henry Adams could be persuaded, conferred with his father, who very much wished that his son "would take the place for five years and then go to Washington again." Gurney was not yet ready to accept the limitation. He therefore turned to the other candidate, Edwin L. Godkin of the *Nation*, from whom he now tried to extract an acceptance. "Henry Adams comes home next week," he told him, "and is likely to be offered an *Assistant* Professorship. I have little doubt a full professorship would tempt him. He would take it, however, only as a stepping-stone in another career." Godkin declined, choosing to follow James Russell Lowell's confidential advice to stick to the *Nation*. "Your leaving it," said Lowell, "would be nothing short of a public calamity." [10]

When Adams disembarked at Boston he little suspected that his rejection of Eliot's offer had not been taken as final. The "question of the professorship" was "sprung upon him in a very troublesome way" he wrote to Gaskell with a certain wry amusement at his own predicament. "Not only the President of the College [C. W. Eliot] and the Dean [E. Gurney] made a very strong personal appeal to me, but my brothers were earnest about it and my father leaned the same way. I hesitated a week, and then yielded. Now I am, I believe, assistant professor of history at Harvard College with a salary of £400 a year,[11] and two hundred students, the oldest in the college, to whom I am to teach medieval history, of which, as you are aware, I am utterly and grossly ignorant . . . I gave the college fair warning of my ignorance, and the answer was that I knew just as much as anyone else in America knew on the subject and I could teach better than anyone that could be had." One of the conditions of the five-year contract was that Henry would relieve Gurney of the *North American* and become its "avowed editor." An agreeable aspect of his new career was that his total income, including the yield of his private investments, would now amount to about $6000. At least he would be comfortable while he rode out the storm which Grant had raised in Washington.

So far as his academic qualifications for the teaching of medieval history were concerned, the appointment seems at first as incongruous as it appeared to him. Hostile critics have attached special significance to the fact that his father was the most distinguished member of the Board of Overseers and that his eldest brother, John Quincy, had been a classmate of the young President. Eliot may also have felt a debt of gratitude to Henry's father, since Charles Francis Adams's refusal of the Presidency of Harvard in 1869 left that office open for Eliot. These influences may have unconsciously operated, but other considerations were far more compelling. Having won its fight against clerical fossilism, the new Harvard administration wanted fresh allies to help impose its reform program upon the university. In his inaugural address Eliot had demanded that history should be taught "by teachers of active, comprehensive and judicial mind," by "young men and men who never grow old." Gurney, a leader of the reform group and one of Eliot's chief advisers, regarded purely technical qualifications as of secondary importance. In his recommendation of Godkin, for instance, he said, "Of course, it is a venture and a serious one to take a man of different training, and who has never taught; but I think so very highly of his taste, his energy, and his personal power that I don't believe that he can be anything but a great success." Adams's youthful predecessor, the positivist John Fiske, had also been a talented nonprofessional historian. As Professor Ephraim Emerton, one of Adams's distinguished pupils, wrote in 1883, "Any 'cultivated gentleman' could teach European history." [12]

By these standards Adams was remarkably well qualified. The new Dean knew him as one of the most frequent and able contributors to the *North American*. One of his weightiest studies in financial and political history had also been welcomed by the *Edinburgh Review*. Gurney likewise knew that Sir Charles Lyell admired his scholarly acumen. Charles Deane, a leader of the Cambridge historical group, had already praised Adams's ability as a historical writer. In addition, he had the confidence of John Gorham Palfrey, the dean of New England historians, who had encouraged his studies for the past nine years. To the readers of the *Nation* and the members of the reform associations of Boston and New York there could be no doubt of "his taste, his energy, and his personal power." As the right-hand man of David A. Wells he had attained a national reputation as one of the bravest publicists of the reform movement. Finally his natural dignity, his quick wit in conversation, his urbanity gained through a half dozen London seasons, would have impressed a far worse judge of men than Dean Gurney.

Although he allowed himself to be persuaded to accept a thoroughly respectable position that would save him from the "Bohemianism" of Washington and that offered a dignified refuge from politics, Henry found it hard to give up the life of the Capital and the independent career that he had so auspiciously begun. The success of his second annual "Session" had made him the ranking censor of Congress; in the fullness of time he might become a power in the press, a distinguished peer of such men as Charles Nordhoff, Murat Halstead, Whitelaw Reid, Manton Marble, Charles Dana, and Edwin L. Godkin. Yielding to the family influence was in truth a surrender, and it was done against his strongest inclinations. There was much however that his practical-minded brother Charles could argue. His two elder brothers were settled in their careers. As railroad commissioner of Massachusetts, Charles could now turn to matters of practical reform. John had his work as a gentleman farmer and perhaps as perennial Democratic candidate for Governor of Massachusetts. Henry was going on thirty-three and his writings had eliminated whatever likelihood there had been of a political position under President Grant. His role as Congressional lobbyist was honorary only and as a journalist he was becoming notorious. He might even be reminded that his grandfather John Quincy Adams had been a Harvard professor preceding his appointment as United States minister to

Russia. At all events, the cumulative pressures of family affection, tradition, and mere prudence exerted their force upon him and he bowed his head, consoling himself with the thought that Harvard was a "hospitable shop which kindly offered me a place at its counter," while "our opponents" were given a chance to overreach themselves. The decision made, he went on to Washington to close up his rooms, and sent a sad little valedictory to Gaskell: "I came on here yesterday, and am very hot, very lonely, and very hard run. I passed an hour today with Secretary Fish, who was very talkative, but there are few of my political friends left in power now, and these few will soon go out. This reconciles me to going away, though I hate Boston and am very fond of Washington."

Chapter Seven

Harvard College Once More
1870–1877

The Academic Rebel

THE DUAL JOB which Henry Adams had undertaken turned out to be considerably more than he had gambled on. For one who had always sworn that he "never would descend to work" the transformation was remarkable. Nine hours in the lecture room each week, even though he was left "absolutely free" to teach what he pleased "between the dates 800–1649," forced him into a paradoxical situation: "I have come back here not so much to teach as to learn," he told Henry Lee Higginson. To keep abreast of his classes he scrambled through "three or four volumes of an evening," many of them "heavy German books." In mock despair he said, "My happy carelessness of life for the last ten years has departed, and I am a regular old cart-horse of the heaviest sort." The first skill which he had to master was the art of improvisation, a task not lacking in perverse entertainment. "There is a pleasing excitement in having to lecture tomorrow on a period of history which I have not even heard of till today." After three months of it, he whimsically reported to Charles Eliot Norton, then sojourning in Italy, "Thus far the only merit of my instruction has been its originality, one hundred youths at any rate have learned facts and theories for which in after life they will hunt the authorities in vain, unless, as I trust, they forget all they have been told." [1] Engrossed with such concerns, Adams looked back incredulously upon the pleasures of English country life, as faithfully reported by Gaskell: "Lord bless me! Do these people really exist, or did I dream it all, after reading Horace Walpole and eating a heavy dinner?"

The odd bits of architectural lore which he had picked up in his tourist pilgrimages found unexpectedly apt use. From Chester to Prague, from Berlin to Palermo, everywhere the Gothic tower of the medieval church had thrust its sword hilt against the horizon. To one brought up on plain meeting houses the attraction had been irresistible. Fortunately, he had read James Fergusson's *History of Architecture* and something of Ruskin when he was in England, encouraged in his study by Francis T. Palgrave. Now he came to the incomparable Viollet le Duc whose essay on military buildings in the Middle Ages excited "a new interest in architecture." Sometimes, when driven to "giving lectures cribbed bodily out of Fergusson and Viollet le Duc," he keenly felt the lack of "a good historical collection of cathedrals in photograph." Often aware of his "massacre" of "the principles of historical art," he was happy that Palgrave was not present to "brandish my ancestral tomahawk over my head and brain me where I sit."

For all his banter, he was beginning to translate the esthetic pleasure which had twice impelled him to toil to the very summit of the Antwerp cathedral into historical sequences not to be found in Baedeker or Murray. The new exploration foreshadowed his lifelong search for the meaning of those Gothic symbols. A poetic chart already existed. Professor James Russell Lowell, his associate on the *North American,* had recently hymned the glory of Chartres cathedral in a long and tumid poem, a poem which must now have taken on a richer meaning to Lowell's former proselyte. Its mood strongly presages that of Adams's own *Mont-Saint-Michel and Chartres.* Lowell, glutted with the perfectness of Grecian architecture, felt his disbelief vanish before the restless Gothic.

> *But ah! this other, this that never ends,*
> *Still climbing, luring fancy still to climb,*
> *As full of morals, half-divined as life . . .*
>
> *Heavy as nightmare, airy-light as fern,*
> *Imagination's very self in stone! . . .*
>
> *I looked, and owned myself a happy Goth.*

"The Cathedral," filled with religious nostalgia and the anguish of the unwilling sceptic, could not help but speak directly to Henry's heart. One day he would respond reverently with a poem of his own: "A Prayer to the Virgin of Chartres."

Old as his historical subjects were they fostered in him no respect for equally old teaching methods. Having been called in to "strengthen the reforming party in the University," he essayed the role with limitless gusto. "The devil is strong in me," he burst out in a letter to Jacob Dolson Cox, "and my rage for reform is leading me into open war with the whole system of teaching. Rebellion is in the blood, somehow or other. I can't get along without a fight." He chafed against a system in which the students' interests appeared to be secondary to those of vested faculty and administrative groups. He resented the burdensomely large classes which forced the teacher to take refuge in lectures and professorial vaudeville. Fortunately Adams could count on being let alone; he therefore proposed as he said to substitute quietly his own notions for those of the college. Iconoclasm was not the whole of virtue, however. He still felt obliged to reassure his fellow-sophisticate of Wenlock Abbey that he was not losing the cynicism proper to a man of the world. "All this is very grand of course. Equally, of course it is probably unmitigated rot." With similar skepticism he remarked to Norton: "A madder choice I can't conceive than that of me to teach medieval history, but they said there was no one better, so I took it as I did the *Review,* to relieve Gurney's difficulty. It amuses me, certainly, but I doubt whether the professorship was established for just that object." [2]

Despite this pose, he responded with all the intellectuality of his nature to the challenge of the classroom. His mind, "robust and virile" rather than "subtle," as one admiring student remembered it, accosted the great figures of history and their interpreters with no more deference than he had paid to Captain John Smith or David Dudley Field. He was at his best in the small class in medieval institutions, an undergraduate seminar of some half-dozen students which met in his well-furnished rooms in Wadsworth House. In order to "give a practical turn" to his young men, he bent the study strongly toward legal institutions, adopting as his chief text Sir Henry Maine's *Ancient Law.* His method, as he explained to Sir Henry, was to encourage his students "to dispute and overthrow if they could every individual proposition in it." He plunged them into the *Germania* of Tacitus, the *Lex Salica,* "translating and commenting every sentence," Maine's *Village Communities,* M'Lennan, Nasse, "and everything else they could lay their hands on, including much Roman law and other stuff." Commonly he assigned a chapter of a book to each student and required a report on it. In the cross-examination that followed, Adams vigorously attacked every doubtful point, not avowing his own view until the end. As a result, teacher and students would "argue in the lecture room by the hour." He aimed at the detachment of the philosopher, a man whose "business it is to reason about life, thought, the soul, and truth, as though he were reasoning about phosphates and square roots" and one whose "pleasure is to work as though he were a small God and immortal and possibly omniscient." As one of his disciples has said, he acted on the a priori assumption that the actions of men followed certain laws. Occasionally the student's reach exceeded his grasp and the assistant professor's immaculately kept ledger showed term papers hopelessly in arrears. The rigorous tests which he set were of a piece with his classroom attitude. Once he remarked, "I could never understand how you fellows got such high marks. I know I wouldn't myself have got sixty per cent on my own examination papers." [3]

At the end of the first year he was ready to concede that "as things go; and as professors run," he had been fairly successful; "but from an absolute point of view," he complained, "I am still nowhere." How long the present excitement would last he could not say; "probably not more than another year, after which I shall be bored." A strenuous summer with the Fortieth Parallel Expedition in Wyoming and Colorado, spent largely in the company of a new-found friend, Clarence King, refreshed him and he resumed his work at Harvard in the autumn of 1871, "feeling much more at home" in his medieval chair. Once again he was "deep in German, working up no end of history," and impatient with his "confounded Review," which was always "claiming attention at unexpected moments." He even "dodged a meeting of politicians at Washington." He saw himself "sinking into provincial professordom with anguish." Nor did he even trouble to tell Gaskell that his brother John had for the fifth time received the Democratic nomination for governor of Massachusetts. He projected a post-graduate course in medieval history for the coming year, not without the sardonic reflection that the labor was "tremendous and the effect nil," but as the lectures were soon "all in arrears" the post-graduate seminar remained unrealized until 1875.

Three separate courses were all that even he could manage.[4] How well he managed these courses was publicly acknowledged in the January 1872 issue of Edward Everett Hale's *Old and New*. The editor declared in his review of Harvard College: "No part of the college course, out of the physical sciences, has been improved so much as the instruction in history. We wish it included more of American history . . . The assistant professor [Henry Adams] has three elective sections, one of seniors and two of juniors, in medieval and modern history; in teaching which he pursues a method, partly of reading, partly of questioning, which finds favor with the members of the committee attending his exercises. 'The teacher,' they observe, 'evidently relies on his power to awake an interest in his questions as the chief incitement to study . . . For a large proportion of the students, the method, in the professor's hands, is an active success.'" He also achieved garbled mention in the British *Men of the Time* in 1872 as having attained "considerable distinction," but to his amusement the British annalist credited him with all of his brother's writings.

The one institution in which the rebel assistant professor attempted no reform was the Monday evening faculty meeting. The courage which he displayed in his magazine articles and the unceremoniousness that made his classes exciting were no match for "thirty twaddlers . . . discussing questions of discipline." He was bored by the "old buffers" and with characteristic violence thought it would be a pleasure "to scalp 'em." As he had no discipline problems, he regularly relieved his ennui during these three-hour sessions by answering his correspondence from the safety of a back seat. "Disorder in his classes would have been unimaginable," according to a reverent student.[5] So far as the faculty minutes show he preserved a perfect silence and after two years quietly withdrew altogether, abandoning the faculty to its solemn thumb twiddling.

He set forth his unconventional views on education in the first long article which he wrote for the *North American* after assuming the editorship, the occasion being thrust upon him in December 1871, when his "staff of contributors suddenly broke down." In "Harvard College" he stated his cardinal principle: "No system of education can be very successful which does not make the scholar its chief object of interest." Granted that the principle might seem a mere truism, it had "rarely been put into practice on any great scale," because it was, "in the daily work of education . . . the most difficult of all principles to act upon."

After enduring so many fatuous discussions of discipline he had reason for saying that "in the great majority of cases the teacher is, in his own eyes, the most important part of a school: the institution or school or system itself ranks next, and the scholar comes last of all." His father, equally rebellious as a member of the Board of Overseers, must have supplied him with materials for the corollary proposition that there were "grown men who look upon a great and influential corporation like Harvard College in the same light as a railway or a banking corporation, with a history which is thoroughly economical, made up of charters, deeds, and statistics of passengers carried, discounts effected, boys educated, and stock watered." Such misconceptions obviously stood in the way of a scientific study of education. The "true historical method" would seek "to know what the student, at any given time, thought of himself, of his studies and his instructors; what his studies and his habits were; how much he knew and how thoroughly; with what spirit he met his work, and with what amount of active aid and sympathy his instructors met him in dealing with his work or his amusements." The best source of information would be student diaries. To illustrate their possible value he quoted at great length from the college diary of John Quincy Adams, part of the enormous journal which his father was then editing. The diary indicated that the relations between students and faculty were not satisfactory and for reasons which still had practical interest: "The duty of giving instruction, and the duty of judging offenses and inflicting punishment, are in their nature discordant, and can never be intrusted to the same hands, without the most serious injury to the usefulness of the instructor."

In the conduct of his courses and in his personal relations with his students Henry Adams forcefully illustrated what he conceived to be "active aid and sympathy." Desiring a study table for his seminar students, he addressed a mock-serious petition "To the Honorable the Corporation of Harvard College," in which "the undersigned Assistant-Professor of History" respectfully requests that the collection of museum oddities hoarded by Sibley, the librarian, should be cast out to make room for it. He got his table. Edward Channing, the historian, recalls that once he encountered Adams in the library and confessed his inability to get started on his assignment. Together they spent an hour gathering books and then Adams put a warning sign on the heap to protect it from the librarian. In this direct fashion he seems to have introduced the hitherto unheard of practice of setting aside books on reserve for the exclusive use of his classes. Such expedients did not satisfy his perfectionist standards for advanced study. Hence, after he moved to his house on the Back Bay in 1873 he regularly held his seminar in the book-walled library, where before an open fire, following dinner, the little group of co-workers explored the monuments of German scholarship, studying the early Germanic codes and the writings of Waitz, von Maurer, Sohm, and other Germans. In addition, they "read and searched many times the whole collection of Anglo-Saxon laws, and ploughed through 25,000 pages of charters and capitularies in Medieval Latin." Adams adopted the utopian practice of acquiring all the important authorities in his field and his shelves bristled with such works as Lindenbrog's *Codex Legum Antiquarum,* Baur's *Geschichte der Christlichen Kirche,* Döllinger's *Kirche und Kirchen, Papstthum und Kirchenstaat, Die Papst-Fabeln des Mittelalters, Der Papst und das Concil,* Dahlmann's *Quellenkunde der Deutschen Geschichte,* Fustel de Coulanges's *La Cité Antique,* and scores of similar works.[6]

His classroom manner, as depicted by Lindsay Swift, was a triumphant union of dignity and frankness. "There was no closing of eyes in slumber when Henry Adams was in command. All was wholly unacademic; no formality, no rigidity, no professional pose" stood between him and his enthusiastic students. He was "a friendly disposed gentleman . . . whose every feature, every line of his body, his clothes, his bearing, his speech were well-bred to a degree." His sardonic and paradoxical comments, uttered in a characteristic nasal and anglicized drawl, made his charges laugh until their sides ached, "but it was the laughter of a club and not a pothouse." He did not allow friendliness to degenerate, however, into undue familiarity so that even in the intimacy of his seminars he smoked his cigar and sipped his vintage sherry serenely aware that such privileges were not for students.[7]

An anecdote of the historian Channing testifies to the resourcefulness and prodigious memory of his one-time master. After a month of investigation on an assigned subject Channing came to class ready to make his report. "Fully primed and quaking in my shoes, I stood up to read my report, when the door opened and in walked President Eliot with a stranger, an Englishman. Adams uncoiled his legs, arose to his full length of about five foot three, greeted the Englishman warmly, gave him a seat, and President Eliot departed. Then Adams without a blush said, 'I will conclude my remarks on the career of Simon de Mountfort,' of whom he had never mentioned a word. But he proceeded with so much learning that the Englishman was amazed and so was I. And Adams forgot all about my lecture, for which I was duly grateful." But mostly his students recall his lively irreverence. Ephraim Emerton once asked, "How were the Popes elected in the eleventh century?" Adams retorted, "Pretty much as it pleased God." A remark in defense of John Adams elicited the unsettling comment, "You know, gentlemen, John Adams was a demagogue." On another occasion when an unwary student asked for an explanation of transubstantiation he snapped, "Good heavens! How should I know? Look it up." For mere facts he always expressed a "profound contempt." He was wont to say with characteristic extravagance "One fact or a thousand, that makes no difference." Doubtless he shocked the diligent plodders by volunteering, "I am a professor of history in Harvard College, but I rejoice that I never remember a date." Those students who recognized beneath this whimsicality an intense concern with relations and sequences in history soon became devoted adherents. Henry Osborn Taylor, another historian who came out of his classes, spoke the feelings of most of his classmates when as a junior he wrote in his diary: "Adams I think I like more than any other instructor; he is so clear-headed and analyzes history so well; also he despises as I do the barren accumulation of knowledge." [8]

If Adams was unusually sympathetic toward his more promising students, it must also be recognized that the authorities were equally considerate of him. As his brother Brooks Adams once remarked, "No man could have been more petted than he at Cambridge." Before his first year was out he had "managed to get into the 'inside ring' . . . the small set of men who control the university" and he could boast that he had things pretty much his own way. His work as teacher and editor brought him into close contact with many of his colleagues—with James Russell Lowell, Chauncey Wright, J. D. Whitney, Raphael Pumpelly, William Dean Howells, and many others, including perhaps Wolcott Gibbs, the chemist. As a member of "The Club," the equivalent for the younger wits of the more august "Saturday Club," he dined convivially with Oliver Wendell Holmes, Jr., John Fiske, Henry James, William James, Thomas S. Perry, Moorfield Storey, John T. Morse, Jr., and other "clever, ambitious young fellows," to use Morse's characterization. He may also have met at this time another young man who was to figure importantly in his later life, John La Farge, appointed tutor at Harvard in 1871. Extremely little contemporary comment has come down, however, on the impression that he made at the time on his colleagues at Harvard. In a letter of William James one catches a single glimpse of him sitting "saturnine and silent" during a luncheon at Gurney's home. This was a highly uncharacteristic attitude. Intimacies with eminent older men, the Titans of the Saturday Club, were slow to form and Adams whose boyhood had been enriched by daily association with Sumner, Palfrey, and Dana felt the lack keenly. Professor Gurney's home was the one "oasis in this wilderness" for him. It was at this oasis that Henry Adams met an intellectual young woman of twenty-eight whom Gurney was tutoring in Greek. She was Marian Hooper, the wealthy and "charming blue" to whom Adams became engaged on February 27, 1872.[9]

From Tea Kettle to Steam Engine

Although Adams knew that his work as assistant-professor of history would form his chief employment, he expected to salvage a great deal from his political journalism in his role of editor of the *North American Review*. At the very beginning he proposed to make the review "a regular organ of our opinion" and thus avoid breaking "entirely from old connections." The plan should have been feasible, for Gurney and Lowell had already opened its pages to such spokesmen of the reform movement as Godkin, Curtis, Shearman, Jenckes, Bradford, and Hodgskins. Henry and his brother Charles had themselves been notable recruits to that company of pioneer muckrakers, Charles's "A Chapter of Erie" having marked him as an authority in the railroad field. Past the meridian of its fame as a quarterly, the *North American* still retained much of its old prestige. Its subscribers formed an influential élite in politics and the learned professions, but their number was almost painfully small, perhaps not more than three or four hundred, although the well-disposed Boston *Literary World* estimated the circulation at the time as twenty-five hundred. Adams took the editorship admittedly "as a last resource since no one else could be found, and at a moment when it was very doubtful whether the publishers would drop it at once." Charles Eliot Norton and James Russell Lowell, as coeditors, had done much to revive the publication, increasing the rate paid to writers of established reputations to five dollars a page. As other magazines were paying ten dollars a page, contributors had still to be persuaded to take part of their pay in glory. It was agreed that Lowell should continue on as associate editor, but his function appears to have been entirely advisory. When Adams entered on the scene the annual deficit had become chronic for want of sufficient advertising. As he soon learned, "Articles enough, and good enough, I can get, but a page of advertisements would offer me more attractions than the cleverest page of criticism I ever saw." [1]

His responsibilities began with the January 1871 issue. As he got to work on it, its political possibilities seemed limitless. Viewing the activities of his former colleagues from the perspective of the editor's chair, he thought that his group might soon return to power; "Two or three years," he estimated, "ought to do it." His correspondence with Jacob Dolson Cox during the harried weeks in which he tried to round up a corps of contributors reveals that there was no lessening of his interest in the tragi-comedy that was playing itself out in Washington. Late in November 1870 he allowed his flock at Harvard to "wait for their historical fodder" while he ran down to New York with Charles to attend a political council of "revenue reformers" and to arrange with booksellers to push the *Review*. Adams, in reporting to Cox, said that he agreed with his brother and Charles Nordhoff that it was still premature to make plans for a political convention which should formally unite the dissident Republicans as they had already been united in Missouri under Carl Schurz.[2]

One of the first persons whom he approached for an article

was his political chief, David A. Wells, whose term in the Treasury Department had expired in July to the evident relief of President Grant. Adams good-naturedly bullied him into acquiescence: "You had better make up your mind to do it at once." Wells, who had already taken up his new duties as Special Commissioner of Taxation of New York, cordially promised to comply. Adams next addressed himself to his friend Cox, proposing as a suitable companion-piece to Wells's contribution, an article on civil service reform. "A statement on your part of the true principles of reform," he said, "based on your experience in office, is very necessary, and to appear with it in my hand at the outset of my editorial career is of decisive consequence to me." Cox, whose tussle with Grant had ended in cold dismissal on October thirtieth, was peculiarly well-placed to present an anatomy of the degrading scuffle for offices. Adams deprecated the fact that his publishers, James Osgood and Company, would not yet let him pay more than five dollars a page, adding: "My own work I need scarcely say is unremunerated. I must make the Review pay for itself before it can be very generous, but if it is successful I will quadruple the rates." [3]

He also tried to draft two other leaders: Samuel J. Tilden and Carl Schurz. His intention, he told Tilden, was to place "on record, by the side of 'The Chapter of Erie,' an account of the Tammany frauds and their history, given by a person whose authority is decisive . . . You alone know the private history of the affair, and have the means of estimating the actors at their proper value." Adams's plea to Senator Carl Schurz touched on another aspect of the whole movement, the rise of the Liberal Republican party in Missouri. Having earlier in the year solicited an article on behalf of Gurney, he now wrote: "I venture again to . . . renew my request . . . I want the public to know, if possible, how far you and your party represent principles which are of national interest; how far free-trade and reform are involved in the result; and what influences have been at work to counteract success . . . I would be glad to extend the range of your influence so far as is in my power." Wells was unable to make the December first deadline; Tilden, preoccupied with New York politics, had to beg off; Schurz was equally busy. Cox saved the issue with an article which Adams found "excellent in both matter and tone." His brother Charles also came to the rescue with a first-rate piece on "The Government and the Railroads"; but the editor's sanguine hopes for a sensational political issue for his debut were sadly deflated. The reviewer for the *Nation* was not well-impressed: "The number before us is, we should say, an average number, though rather below than above average." When Adams wrote to Norton for an article for the April issue, his tone was distinctly chastened: "Of my success with the *Review* I am far from sure." [4]

Patently, he would have to do better if the magazine was to regain lost ground, for "without literary success," as Adams well knew, "there is no chance of reaching up to" the financial problem at all. Fortunately, a suitable restorative already brewing was soon to justify Lowell's approving comment that Adams was out to make the old teakettle think it was a steam engine. [5] "The New York Gold Conspiracy," which had at last been dropped into the muzzle of the *Westminster Review*, exploded against its targets so brilliantly as to satisfy all of its author's wishes for advertisement. From James McHenry, the financier whose railroad speculations had once been the talk of the Legation when Henry Adams was there, [6] came the rash threat of a libel suit. Eagerly sensing the possibilities of an international row, Adams wrote to London that he would welcome a court test of his article. McHenry, who was annoyed at being linked with Grant's notorious brother-in-law, subsided into unappeased silence. David Dudley Field, more tender of his dignity, wrote a long and indignant letter to the *Westminster* on February 13, defending his retainer by Fisk and Gould.

Late in February 1871 Adams made another political visit to New York where he learned that the "men of Erie" had discovered the authorship of the offending article. He foresaw that New York would "soon be too hot" for him. Cyrus Field, the promoter of the Atlantic cable, with whom Adams had become acquainted in England, called early one morning to remonstrate with him for the attack upon his brother; but Adams prudently kept out of sight and avoided a possible altercation. He was not cowed by the uproar, however, though he knew that the Erie crowd might give him a taste of the legal hoodlumism which had put Sam Bowles of the Springfield *Republican* in jail for a day. At that, jail might be advertisement *in excelsis*. To the grandson of John Qunicy Adams the prospect of "a very lively scrimmage in which some one will be hurt" called for the application of a single principle of strategy: an immediate offensive. With his brother Charles, now a veteran of the Erie War, and Albert Stickney, the New York attorney who represented an aggrieved claimant in the Erie proceedings, Henry Adams began "concocting our new attack" for the April number of the *Review*. Not one, but two articles spread the scandal of Erie before its readers, Charles's lead article, "An Erie Raid," making a frontal attack upon the management, and Stickney's "Lawyer and Client" managing the flank with an impeachment of Field's conduct as counsel for Fisk.

One of the most spirited of Field's defenders was the *Legal Gazette* of Philadelphia which devoted six columns of its May 12 issue to deploring the unfairness and lack of patriotism of a certain "Mr. H. B. Adams of Boston," who "thus sought *abroad* to destroy" the reputation of so eminent a lawyer as David Dudley Field. Later, the *Gazette* spokesman was "happy to know" that his remarks had received "general approbation," especially in the *Real Estate Register* of Pittsburgh which had characterized Adams's conduct in terms "we now forbear to quote." The spokesman's sense of outrage was increased when he learned the identity of the offender. "If the sons of our public men abroad can thus dishonor in foreign eyes our honored citizens at home, it behooves us to see what sort of sons our public men possess, before we send them far from home. And our late minister, whose son thus prostitutes *his* well deserved honor in a way so open to remark, should see that such things are done no more." This was almost better than being called a "begonia" by Senator Howe.

Somewhat disconcerted by Field's reply to Adams's accusation, Dr. John Chapman, the editor of the *Westminster Review*, wrote two letters to the author about it. Adams's explanation evidently satisfied Dr. Chapman, for the July issue briefly stated that "except in one or two particulars, which are in no way essential, we have found no occasion to modify or retract any of the statements made in the article." Besides, the publication in the New York *Tribune* of the correspondence between Adams's friend General Francis Barlow and David Dudley Field, in which Barlow elaborated the charges of malpractice already brought by Sam Bowles had so well-aired the alleged misconduct of Field that further rebuttal was unnecessary. Adams even felt a twinge of remorse for having unleashed the April *North American* upon the unhappy jurisconsult. There could be no question that the journalist reformers of the "new party" had shown their power to educate public opinion.

Adams's part in the campaign against the Lords of Erie came to an end in September 1871 with the joint-publication by himself and his brother of *Chapters of Erie and other Essays.* Here in permanent form were reprinted the first mature fruits of their collaboration. Henry contributed "Captain John Smith," "The Bank of England Restriction," "British Finance in 1816," "The Legal Tender Act," and "The New York Gold Conspir-

acy." Charles made up his share with three of his railroad articles. In a long footnote to "The Gold Conspiracy" Henry
adverted to his charge that Field had influence over certain
New York judges. "Field's modest denial," it seemed to him,
was "hardly calculated to serve as a final answer." In any case
the Barlow letters in the *Tribune* now made the point incon-
sequential.

The reprinted articles give striking evidence of Adams's pre-
occupation with questions of literary style. The following
parallel passages from "Captain John Smith" exemplify the
wholesome surgery practised on all five articles.

1867	*1871*
The first idea that occurs . . .	*The instant result . . .*
It is to be hoped that this feeling	*Perhaps this feeling . . .*
In the enthusiasm which her act	*In the general enthusiasm,*
has called out, language . . .	*language . . .*
The most careless reader will see	*One sees at a glance . . .*
at a glance . . .	
It is not however to be supposed	*The argument . . .*
that the argument . . .	
. . . are falsehoods of an ef-	*. . . are falsehoods of a*
frontery seldom equalled in mod-	*rare effrontery.*
ern times . . .	

A few of the changes, however, have more than a stylistic
interest; they reflect the sobering effect of second thoughts on
some of his more vituperative phrases. Though he did not
alter the malevolent portraits of Fisk and Gould in "The Gold
Conspiracy," he discreetly withdrew the lie direct against Fisk.
Spaulding, whom he had pursued into the columns of the *Nation,* fared slightly better in the reprint of the "Legal Tender
Act." The accusation that he had "perverted legislation into an
instrument of fraud, and is proud of it," for example, was

wholly deleted, and other harsh passages were softened. The sentence "To this suggestion Mr. Spaulding made a response which must be quoted word for word, since it probably contains the most extraordinary revelation of ignorance and folly that has been offered" became the innocuous "To this suggestion Mr. Spaulding made the following response."

The American reviews of *Chapters of Erie* impressed Adams as "highly civil"; one of them, in the *Literary World,* referred to the essays as "perhaps the best history of some of the most wonderful financial events." But Adams particularly coveted an English review. Thanks to the "laudable efforts" of Gaskell such a notice appeared in the London *Saturday Review,* on December 30, but it gave the lion's share of the praise to Charles's railroad articles. With true British complacency, the reviewer scornfully concluded that the volume gave "most instructive evidence on the practical working of extreme democratic theories." Gaskell made suitable amends by himself writing a "judicious and handsome" notice of the work for the *Saturday Review* for January 13, 1872.

Almost simultaneously one of Adams's English friends, Francis Lawley, took to violently "puffing" the *North American Review* in the London *Telegraph,* describing it as having "sprung into existence." "A fact," said Adams, "which sounds queerly in this benighted land, where the periodical has been hitherto considered as a species of medieval relic, handed down as a sacred trust from the times of our remotest ancestors." In spite of all this advertisement, however, he was unable to realize his hopes of making the magazine a truly effective political "organ." He had kept after Wells for "an authoritative announcement of our proper policy" and commanded him to "make it as strong as you can." Wells delivered his "The Meaning of Revenue Reform" in time for the July issue, but his counsel was very temperately phrased and the exposition was clouded with statistics. Carl Schurz still continued to elude him, and as a result Adams could fairly complain that the July issue was "a very dull number."

In October Charles Nordhoff stirred the political pot with "The Misgovernment of New York," but its savor was somewhat hidden among several historical and scientific papers. By this time, Adams was making the desperate suggestion to Schurz that he submit at least a rough draft of an article which could be worked up into a manuscript and returned to him for "correction and improvement," but Schurz cannily kept himself free from entanglements. Once more Adams turned to Cox; "We are rapidly subsiding into political indifferentism," he warned. "In view of the coming Presidential election . . . a pronunciamento would attract interest and rouse action." The appeal failed. Again his brother came to the rescue, this time with an anonymous article on "The Butler Canvass" in which after some savage comments on the greenback advocate Benjamin F. Butler, he turned to the issue which had been injected into the campaign by the Labor Reform Party. He urged Massachusetts workingmen not to be misled by the socialistic ideas of self-appointed labor leaders. The remedies of the Commune, recently liquidated in Paris, and the Workingmen's Internationale were alien to Massachusetts; the conflict of labor and capital could be solved, he said, through traditional American institutions. As if to underscore that lesson, Henry Adams printed Gryzanowski's unsympathetic study "On the International Workingmen's Association" in the following issue, for April, 1872.[7]

After the turn of the year, Adams's editorial interest in domestic politics lapsed, not to be revived for two years. Yet never was the outlook so bright for his friends and for the members of his family. In January the Missouri Liberal Republican convention, under the leadership of Carl Schurz, issued a call for a national meeting in Cincinnati on May 1; the Massachusetts group added its voice on April 17. A movement was soon afoot to draft Charles Francis Adams who had just been appointed to head the American delegation at the Geneva Arbitration for the settlement of the "Alabama" claims. By the date of the Cincinnati convention his nomination for the Presidency seemed fairly certain, in spite of the known opposition of Charles Sumner who remained aloof from the proceedings.[8]

Henry's brother Charles actively worked with such colleagues as Atkinson, Bowles, Halstead, Watterson, White, and Wells, but Henry had begun to feel indifferent for he was now thinking of marriage. "I look with great equanimity upon the event of the choice falling on some other man," he wrote to Gaskell. "Meanwhile, the fight makes a useful counter-irritant to love. My fiancée, like most women, is desperately ambitious and wants to be daughter-in-law to a President more than I want to be a President's son." On the first five ballots Charles Francis Adams led his nearest opponent, Horace Greeley, by a wide though diminishing margin. On the sixth ballot an inexplicable stampede—shrewdly engineered by professional politicians—swept the hall and at the end of the wild uproar, the vote stood: Greeley 332, Adams 324. The fiasco did not upset Henry, as he believed that his father had "come out of the fight very strong and sound," having behaved with the most dignified reserve toward the overtures which had been made to him. As to the successful rival: "If the Gods insist on making Mr. Greeley our President, I give up." As it turned out, the gods sided with Ulysses S. Grant, whose misdemeanors were at least predictable.

Editorial Principles

Adams's violence against the *status quo* in teaching was matched by his violent unconventionality as an editor. If he had dared he would have flown a skull and crossbones on the masthead of the quarterly. He wanted articles that would carry out his favorite intellectual maneuver of smashing things generally. "If you care to write thirty pages of abuse of people and houses in England," he cajoled Gaskell, "send the manuscript to me, and if you are abusive enough you shall have £20." He promised Norton to "put more energy into the literary notices," and he undertook to do so with his customary unrestraint: "Stand on your head and spit at someone," he admonished his English friend. "Rake up a heap of old family scandals, no one will know who did it . . . Show a surprising familiarity with English affairs, and create a circulation for me there." The "peculiarities" of Labouchère's current bestseller, *Diary of a Besieged Resident in Paris*, entitled it to "a gentle roasting"; Hayward's *Essays* might similarly be reviewed with pleasure, "especially if you did it viciously enough." But Gaskell proved a disappointment; his reviews always came out fairly innocuous.

In Adams's eyes the editor was not a passive intermediary between author and publisher, but rather an ex officio collaborator with full powers of revision. As a result he left a reputation of being "dictatorial." [1] As a novice writer for the *North American* he had expected—even solicited—Charles Eliot Norton to correct his manuscripts; he now assumed an equal humility in his contributors. On one occasion, receiving an article by "an eminent local historian and antiquary," he directed his young assistant, Henry Cabot Lodge, to "go over it and strike out all superfluous words"; on another occasion he directed him to say to a writer, "The editor reserves to himself in all cases, and rigidly exercises, the right to strike out or modify expressions which he deems too strong." Adams attained his autocratic best in the conduct of the department of "Critical Notices."

Once, however, he caught a Tartar and gave considerable embarrassment to his publisher, James R. Osgood and Company. As a book publisher, Osgood brought out Bayard Taylor's poetical translation of *Faust* and Adams arranged to have it reviewed. The review never went beyond the page proof stage, for Adams, in cutting it down from twenty-five pages to ten and drastically altering its character, infuriated Taylor, who promptly took his case to Osgood. "I have worked for years in the interest of the *North American Review* urging people to support it as the best of our periodicals," he fumed, "but if the *North American Review* is to represent Adams with his tastes and prejudices only, it is time to take another view of the matter . . . If there is to be an exclusive Boston circle created in our literature, the sooner the rest of us know it the better." When Adams had finished tampering with the review it unsparingly attacked Taylor's principle of translation: "To copy slavishly the measure of every line, and the very order of the rhymes, is a devotion to principle which approaches fanaticism." In the translation of the Second Part, Taylor had allowed himself, it was true, more latitude, but Adams conjectured that in so deviating from his principle the translator had "yielded not so much to the impulses of the poet as to a very natural and inevitable sense of fatigue"; hence for the most part he found Taylor's rendering of key passages inferior to the efforts of earlier translators. On a proof sheet Adams meticulously inscribed the following obituary: "This notice, written originally by a strong admirer of Mr. Taylor, but much changed by me in tone, led to a protest from the author, and a request from Mr. Osgood that the notice should be suppressed. Which was done." [2]

Privately Adams was willing to concede that "he was no great hand at poetical criticism" and in his first try at a review of his friend Palgrave's *Lyrical Poems* he had to give up; but like most persons he knew his own preferences. Brought up on the eighteenth century writers, his tastes were distinctly classical. Hence he altered a review of Chatterton with the remark that "a generation which had Gray and Collins for models, and produced Burns" could not fairly be called "dust and ashes." Though "everyone now snubs the last century," he still believed, "Pope a poet, and Gray, too, and Cowper, and Goldsmith." When he finally returned to Palgrave to review *A Lyme Garland* it was Palgrave's classicism that attracted him: "Palgrave's highest ambition would be to offer the classical beauty of Greek form to an age and generation which has hardly a notion of form at all; which loves roughness and extravagance for its own sake." [3]

Adams took as his special province for review the writings of contemporary English historians of early law and institutions. As a convert to the German historical method, he reproached English scholars for their inferiority. In his judgment English legal history had suffered from the intellectual narrowness of the Inns of Court because the English lawyer, unlike his German counterpart, was trained exclusively to be an advocate and not a philosophical jurist. "It is the misfortune of England," Adams said in the course of a highly commendatory review of the German Rudolph Sohm's *Lex Salica*, "that she has never yet had a scientific historical school." And when he reviewed Professor George W. Kitchin's *History of France* he found no reason to abate his harsh view. He had hoped to see "better fruit from the new English school" represented by that Oxford Don. It appeared, therefore, that the early history of English private law was "destined to remain untouched until Germany forced England into scholarship." [4]

In dealing with the English writers he strove to use language at least as slashing as that common to the English quarterlies with whose amenities he had long been familiar. Thus he obliterated Kitchin's chapter on "Feudalism and Chivalry" as "downright rubbish" and attacked the eminent Edward Augustus Freeman, historian of the Norman Conquest, for writing passages "altogether bungling and inartistic in effect." The perverse rationale of some of these reviews is explained in a

letter to Sir Henry Maine, written just after he had read Maine's
Early Institutions. The book had pleased him. "Of course," he
added, "I shall attack it ferociously. There is a delightful want
of responsibility in writing in this country on such subjects,
a feeling that no one will read it or care for what is said, if
they do; and this stimulates the writer to the most reckless
tours de force by way of criticism." On occasion he did feel
some qualms about his method. After he had published a few
sharp comments on George Bancroft's literary offenses, he hur-
riedly abased himself before the old man: "So far as views of
the general subject are concerned, your opinion is worth so
much more than mine that I have not hesitated to express mine
freely, on the general ground that it is a matter of small consequence whether they are expressed or not." [5]

From his seminar he borrowed the technique of questioning
all authority, of probing all assumptions, of tracking effects to
their ultimate causes. Here he insinuated doubts; there suggested alternative possibilities, fresh applications, wider extensions of principles. Maine's treatment of the Brehon Celtic law
he thought defective because it seemed to assume the archaic
character of that law. It must be proved to be archaic, he
argued, before it could be "turned to the uses of comparative
jurisprudence." Fustel de Coulanges in his *Ancient City* had
grasped "an idea full of promise" in showing that "the institutions of archaic Rome grew from, and were in every fibre permeated with, one great religious idea." But now that "several
years" have passed since the book was originally issued "the
student should begin to ask what further use can be made of
the historical principle developed in it." A newcomer in the
field, he nonetheless chided his colleagues for their ignorance
and their neglect of duty: "The early history of law has never
been written." Indeed, the student of English history, after
reading Mariette Bey's model study of early Egyptian institutions, could easily see "that no really thorough acquaintance

with his subject matter is possible without tracing the stream of legal institutions back through the German hundred, as well as through the Roman city to its Aryan source."

He counselled his English confreres to be bold; but he also cautioned them against an excess of that virtue. Freeman was excessively English in his hostility to theories;[6] Maine, however, erred on the other side. It might be possible, Adams conceded, to reconstruct the archaic Indo-European jurisprudence, as the archaic Indo-European language has been reconstructed in its successive stages, but each of Maine's theories would require "two or three volumes" of preliminary scholarship, and science was far from this point. As if to exorcize his own strong inclinations he interpolated a caveat against the system builders: "The mania for producing philosophical systems of social development, like that of August Comte; the temptation to make such a system symmetrical and to pass every individual human being through every phase which has left a trace behind it; the fervor with which each new investigator presses his own historical novelty;—all this is merely the symptom of advancing knowledge, but it has little intrinsic value."

Such was the counsel of prudence and practicality. Yet, if at the moment young Henry Adams was no system builder, he was preoccupied with theories of history out of which systems are built and he found himself strongly attracted to the "science of comparative archaic jurisprudence" because it promised to forge some of the most needed links in "the chain of human development."

Running through all his incisive and opinionated reviews, no matter what their subject, was an almost obsessive interest in effectiveness of expression. No writer whose style lacked grace and vigor escaped censure. He lectured even the great constitutional historian Bishop Stubbs: "The historian must be an artist. He must know how to develop the leading ideas of the subject he has chosen, how to keep the thread of the narrative always in hand, how to subordinate details, and how to accentuate principles." Of such an art, "of late superbly developed" in Germany, Stubbs was "altogether innocent." Not that the Germans were wholly faultless: von Maurer's essays he thought "flighty and diffuse," not to be set beside "the books of Sohm and Brunner, Thudichum, Heusler, and so many other German scholars." He was equally unable to forbear taking his friend Bancroft to task for "the inevitable peculiarities of his style" especially his tendency to fall into his "old excursive ways."

Among the English historians, only John Richard Green delighted his sense of literary form. "Never has the popular style of historical writing been raised to so high a standard as in Mr. Green's work." The *Short History* attempted an impossible task of synthesis, an attempt which the advance of historiography would steadily undermine, yet "nothing can ever take away its calmness of judgment, its elevation of tone, or its beauty of style." Green's omission of the literary figures of the Augustan and Romantic periods seemed to him, however, a blemish upon an otherwise artistic narrative, and he construed it as evidence of "declining interest . . . haste . . . natural fatigue." The Augustans, he remarked defensively, may be little read "but they are history for all that, and Mr. Palgrave's remark is worth remembering, that 'an intelligent reader will find the influence of Newton as markedly in the poems of Pope, as of Elizabeth in the plays of Shakespeare.'" He would probably have laughed had he known the blemish was deliberate. Green had simply run out of space and his publisher was adamant. Later English literature had been economized out of the manuscript.[7]

Adams indulged his passion for stylistic criticism even in his private reading and a burning trickle of comment would often run down the margin of his book. In Edmund Burke's *Abridgement of English History*, for example, he jotted down such editorial memoranda as: "Strange assertion for a contemporary of Gibbon"; "Surely this was always a colloquial phrase!"; "This is very droll"; "Curious want of critical capacity"; "All this is slovenly and incorrect"; "This is bad grammar, bad style, and bad sense"; and much more of the same, as if in imagination he were revising the text before him. Even more thorough was his editing of the "Speech on American Taxation," in which he corrected the grammatical errors, rephrased awkward expressions, and questioned the taste of many of the epithets, with a thoroughness worthy of Benjamin Franklin's similar self-discipline. To incite himself to greater perfection he once proposed to young Lodge that each write a review of von Holst's *History* "and then we will take what is best out of each and roll them into one . . . I think it will give us both a spur of emulation, so that we shall do the most perfect work we can." Rigorous toward himself, Adams was equally rigorous toward others. "He sent me to Swift," Lodge tells us, "to study simplicity of style as well as force and energy of expression . . . My first article was a critical notice of Baxman's *History of the Popes*. I rewrote it eight times before it passed muster." [8]

Wedding Journey of a Scholar

With marriage impending, Henry Adams now added a third term to his usual complaint of overwork: "What with teaching, editing, and marrying, I am a pretty well-occupied man." Restless of spirit, he had been more than usually elusive, instinctively recoiling from the yoke which his friends persisted in accepting gracefully. He had humorously pretended that he who married must "be an unmitigated and immitigable ignoramus and ruffian." Now at thirty-four his defensive bearishness merged into the pose of scholarly detachment as he contemplated matrimony: "Socially the match is supposed to be unexceptionable," ran his comment to Gaskell.[1] "She is certainly not handsome; nor would she be quite called plain, I think

. . . She talks garrulously, but on the whole pretty sensibly. She is very open to instruction. *We* shall improve her. She dresses badly. She decidedly has humor and will appreciate our *wit*. She has money enough to be quite independent. She rules me as only American women rule men, and I cower before her. Lord! how she would lash me if she read the above description of her!" He realized he could not gracefully play the role of swain, but there was still a good deal of romantic sentiment in his make-up. The young idealist who had read Edmund Quincy's *Wensley* in *Putnam's* magazine and had written in the *Harvard Magazine* about "an angel in muslin" lived on in him. By coincidence, his friend William Dean Howells had just published *A Wedding Journey;* Adams decided to review it himself. Admiringly, he spoke of the "delicacy of touch" which argued feminine assistance. "The book is essentially a lover's book. It deserves to be among the first of the gifts which follow or precede the marriage offer . . . If it can throw over the average bridal couple some reflection of its own refinement and taste, it will prove itself a valuable assistant to American civilization." He asserted that "our descendants will find nowhere so faithful and pleasing a picture of our American existence." For once, the iconoclast wrote like a cooing dove.

His stint for the April *North American* had also included two other brief notices, one of them being of somewhat curious biographical interest. This was a review of Clarence King's *Mountaineering in the Sierra Nevada.* Oddly enough it gave no hint of the arduous weeks spent in the saddle as King's companion and its impersonality comes as a shock after the account of their dramatic meeting in *The Education.* He had no inkling that the collection of realistic sketches of life in the high mountains would one day be valued as a minor American classic. All he saw was that the book, "though agreeable reading enough, is but a trifle."

Adams's political interests yielded not only to the claims of romance, but also to the increasing attractions of medieval scholarship. Before his first year of teaching was up he came to feel himself "all behindhand in the gossip of my trade" and he laid plans to "go over next year" to improve his acquaintance with John Richard Green and to consult with Stubbs at Oxford, meanwhile "solemnly" reading their works in anticipation of his visit. The prospect of an extended wedding journey to the Eastern Mediterranean promised an opportunity for "medieval work in France and Germany" as well. What Adams thought of the historical profession on the eve of his scholar's holiday appears in a letter of advice which he sent to Henry Cabot Lodge, who had been one of his most promising students. "There is only one way to look at life, and that is the practical way . . . The question is whether the historico-literary line is practically worth following, not whether it will amuse or improve you. Can you make it *pay*? either in money, reputation, or any other solid value. Now if you will think for a moment of the most respectable and respected products of our town of Boston, I think you will see at once that this profession does pay. No one has done better and won more in any business or pursuit, than has been acquired by men like Prescott, Motley, Frank Parkman, Bancroft, and so on in historical writing; none of them men of extraordinary gifts . . . What they did can be done by others . . . With it comes social dignity, European reputation, and a foreign mission to close."

In preparation for the trip Adams resigned the editorship. At Howell's urging, the publisher appointed young Thomas Sergeant Perry to the "sub-editorship" to keep the magazine going. But the burden of getting out the next number fell upon Howells, because, as he protested, Adams and his associate Lowell "sailed before any permanent provision for the editorship" could be made. Lowell took formal leave of absence from Harvard by offering his resignation in writing; the faculty minutes make no note whatever of Adams's leave. At this unpropitious moment Whitelaw Reid, editor of the New York *Tribune,* invited Adams to join his staff. Pleased by the recognition, Adams forebore any ironical comment. "If you still feel like calling on me when I return, I may succumb to the temptation though I am tied here by the leg." The fateful temptation was not again to be offered. His wedding took place very privately on June 27, one of Adams's churchman book reviewers performing the ceremony. In the party that left for England on July 9 were the Lowells, Francis Parkman, and John Holmes, brother of Dr. Oliver Wendell Holmes. As usual the voyage made Adams very seasick: "Worse than ever," he recorded in a joint travel letter written aboard ship. "Deadly sick and a calm sea. Wish I were to hum. Wish I were dead! Wish I'd never been born!" Though he was to become one of the most widely traveled of men, he never succeeded in making his peace with Ocean.[2]

In order to delay their arrival in Berlin until September when more persons would be accessible, they proceeded first to Geneva, where Charles Francis Adams was engaged in the final deliberations over the "Alabama" claims. After a short stay they pushed on to Berlin, finished Henry's business there in a week, and then headed south for Italy and the Nile. In Egypt he busied himself with his new photographic equipment and tirelessly studied the vestiges of a civilization so ancient as to make his own field of study seem modern. Marian Adams could not share his intensity of interest, being, as she confessed, "painfully wanting in enthusiasm," for what she did not understand, but like a sensible wife she did not stand in his way. In the long days on the Nile her scholar-husband went through "a library of books" carted from Germany. All trace of cynicism vanished from his manner. Marian reported to her father that "Henry is utterly devoted and tender." Henry on his part addressed an incongruously respectful letter to his father-in-law in which he told that Marian had "gained flesh and strength, so that she is in better condition, I think, than I have ever known her to be." It was an auspicious beginning of that halcyon period at whose sudden tragic close he wrote, "For twelve years I had everything I most wanted on earth." [3]

Wherever they went he satisfied his long-standing plan to meet the important European historians. On the way to Geneva he had stopped off at Bonn to talk shop with Professor Heinrich von Sybel, editor of the *Historische Zeitschrift* and historian of the early Crusades, and found him "kind and communicative." In Berlin the American Minister, George Bancroft, who was a cousin of Marian Adams, gave two dinner parties for them to which he summoned the most notable German scholars in history and public law. Aged Leopold von Ranke, the first of modern historians, was not among the guests on either occasion, but his chief disciple, von Sybel, was an adequate spokesman for the scientific method in history. Mommsen represented the domain of Roman law and history; Ernst Curtius, the archaeology and history of Greece, and Heinrich Rudolph von Gneist, the history of German legal institutions. Another guest, George H. Pertz, venerable editor of the *Monumenta Germaniae Historica,* whose long career of scholarship and politics was then drawing to a close, was especially "cordial and kind." One evening Adams took tea with the learned old doctor and was shown some of the important historical materials Pertz was then working on.

Among these distinguished authorities Adams felt most drawn, however, to von Gneist and he "pumped" him diligently all one evening.[4] As the foremost European student of English constitutional and administrative law, von Gneist had most to offer him with respect to early English constitutional history, a subject which Adams had begun to teach—and study intensively—only a little more than a year before. Von Gneist's *Das englische Verwaltungsrecht* very likely had been one of those "fearful German books" with which Adams struggled during the previous winter. In the German master he could recognize a kindred ambition, for von Gneist in spite of his immense erudition was intensely practical in his approach to legal studies. Tens of thousands of students flocked to his lectures, yet he was equally celebrated as one of Prussia's greatest Liberal statesmen.

En route across Europe Adams continued "reading history in German." Fortunately his wife shared his interest, and they enlivened their wedding journey by reading together in German Schiller's *Thirty Years War*. In Florence the versatile German scholar Dr. Ernst Gryzanowski called upon them and so drenched his hosts with Hegelian metaphysics that Mrs. Adams thought it "enough to last for a long while." Gryzanowski had been one of Godkin's finds for the *Nation,* a find soon shared with the *North American*. It was Gryzanowski who through the variety and frequency of his contributions to the *North American* came to be esteemed by Adams as his "best contributor." [5]

By May, 1873, Adams was again back in London, enjoying as a "wedding present" the use of his friend Gaskell's town house in Park Lane. James Russell Lowell came over from Bruges in June to stay with him for a few days "until made a Doctor at Oxford," later sending back a long whimsical bread and butter letter in rhyme.[6] Adams had already made a professional visit to Oxford, where he "saw all the men I expected to see— Stubbs, Burrows, etc.—and a number I did not expect to see— as Sir Henry Mayne [Maine] and Laing of Corpus." To arm himself for his interview with William Stubbs he first "inspected the early English manuscripts in the Bodleian." Stubbs, the regius professor of history, was then at work on his *Constitutional* History of England, a project of especial interest to Adams after his interview with von Gneist. He was of course thoroughly familiar with Stubb's *Documents Illustrative of the Constitutional History of England,* as it was required reading

in his Harvard course on English history. The meeting with Sir
Henry Maine, who was professor of historical and comparative
jurisprudence, gave Adams the opportunity to exchange views
with the author of the two chief staples in his undergraduate
seminar in medieval institutions, *Ancient Law* and *Village
Communities*. He also dined with Benjamin Jowett, regius professor of Greek and Master of Balliol, and lunched with Montague Bernard, the professor of international law and diplomacy. Of the Cambridge historians he seems to have met only
Sir John Seeley, professor of modern history. Yet amidst all
these giants of his profession Adams was not tremendously
impressed. "The English universities run too much into money
and social distinctions," he reflected. "The spirit is better in
ours."

Some of the values and results of the year abroad are suggested in the letters which Adams wrote to Lodge who desired
guidance for his graduate studies. "The first step," counselled
Adams, "seems to me to be to familiarize one's mind with
thoroughly good work, to master the scientific method, and to
adopt the rigid principle of subordinating everything to perfect
thoroughness of study . . . Of course our own law and institutions are what we aim at, and we only take German institutions so far as they throw light on English affairs . . . It matters very little what line you take provided you can catch the
tail of an idea to develop with solid reasoning and thorough
knowledge . . . America or Europe, our own century or prehistoric time, are all alike to the historian if he can only find
out what men are and have been driving at, consciously or unconsciously." From London he added "I have this year been
engaged in investigating and accumulating notes upon some
points of early German law, out of which I expect in time to
make a pamphlet or small book. If you like, I will put these
notes in your hands next term, and we will proceed to work
the subject up together." The seminar was now to have its

work cut out for it in "German, French, Latin, and Anglo-Saxon."

In London the former private secretary garnered a rich social harvest from his earlier residence in the city to which he now returned as a well-known editor and writer. His letters and those of his wife chronicle the renewal of old friendships and the making of new ones and crackle with the gossip of incessant entertaining. The beadroll of their pages includes Sir Charles Trevelyan, the India statesman and historian; Dean Arthur Stanley, one-time professor of ecclesiastical history at Oxford and now Dean of Westminster; Frederick Locker the poet (not yet married to Hannah Lampson whose surname he took); Francis Palgrave, the poet and art critic with whom he was on the friendliest terms; Thomas Woolner, the pre-Raphaelite sculptor; Stopford Brooke, the churchman and man of letters; and Lord Romilly, the Master of the Rolls. He paid his respects as well to the two great statesmen who had stood by his father during the darkest days of the Civil War: William E. Forster and John Bright. An encounter with Robert Browning, another one of the Legation-elect, left Mrs. Adams unimpressed. "He's not exciting," she wrote of the ageing poet. "I don't hanker to see him again."

After all his anticipation Adams appears to have missed meeting with John Richard Green, the brilliant friend of Stubbs and Freeman and youngest member of the Oxford school of historians. He did have the chance however to compare notes with young James Bryce, whose professorship in Civil Law at Oxford was as recent as his own. To the Park Lane house there also came the editor and critic Leslie Stephen and old friends of the Legation circle, like Thomas Hughes. Most welcome perhaps was Gaskell's cousin, the affable young M.P., Sir Robert Cunliffe, one of Adams's dearest friends. As the time of departure drew closer Henry and Marian Adams became increasingly aware of the attractions of life in a world capital. "I like giving dinners in such a big society—one can get more variety of material than in Boston," she told her father. With far more reason than ever before Henry could repeat his old ejaculation: "How much of this sort of thing could one do in Boston!" The quiet provincialism of Quincy, Boston, and Cambridge would be even harder to endure henceforward.

Work and Plans

The two travelers got back to Boston early in August 1873 "bobbing up on this side of the ocean, like a couple of enthusiastic soap bubbles" and beguiling their friends with extravagant tales of their travels. Adams needed to be gay for he would have to call on all his resources of sardonic comedy to cope with a schedule now demanding "twelve hours a week in the lecture room." [1] Fortunately, he would not have to reshoulder the editorship of the *North American* until after January 1, 1874, by which time the January issue would be safely out of the way. Moreover, the newly purchased house at 91 Marlborough Street, Boston, would provide a refuge three or four days a week at a safe distance from his hundred students.

Washington now attracted him more than ever as the nearest equivalent to London, and in February he again set foot upon his "native asphalt" to enjoy surcease from the "ways of Boston" and to smile "contemptuously on men in high place." Official life as a career repelled him as always, although its workings and its society fascinated him. "To be a free lance and to have the press to work in" was his "ideal of perfect happiness, at least so far as perfect happiness is to be found in a career." Such reflections, however, were hardly more than a fond backward glance at those Washington days when he had been wound up in a coil of political conspiracy.

Immediately after reassuming the editorship of the *North American* Adams asked Henry Cabot Lodge, who was then twenty-four, to act as assistant editor; the young man leaped at the chance. "Nothing has ever come to me which gave me so much joy as that offer," he later wrote.[2] Under the division of labor agreed upon, Adams retained active general direction of the magazine and Lodge was responsible for getting out each issue once the manuscript was turned over to him. By this means the editor freed himself from much time-consuming drudgery; in exchange he directed Lodge's literary apprenticeship with unexampled patience. Once he threatened to make Lodge edit an entire issue; but apparently on second thought he concluded that such a task might be an imposition upon the young man.

Despite the valiant aid which Lodge gave him, Adams continued to suffer from countless demands upon his energies, especially after he began to work his way back into the political councils of the Liberal Republican remnant in Massachusetts. As time went by the "dreary waste of examination books and Division Returns" became more rather than less onerous; and the petty exactions of teaching intruded unpleasantly upon larger work. Ambitious students with "no minds" had to be reconciled to their lot; more richly endowed students had to be rescued from undue humility. If his staff of contributors were not "all behaving like the devil," he was busy conciliating men like Parkman or Gamaliel Bradford who offered articles that could not easily be accommodated, or peremptorily demanding articles as he did of Howells ("Can you fix a day for letting me have that article?"), or asking him to stand by with one "in case one of my men, as it is probable, disappoints me." Occasionally, however, aid came from unexpected sources, as when Ralph Waldo Emerson solicited him to take an article by the Scottish philosopher James H. Stirling denouncing Buckle's philosophy of history. But no matter what the state of his relations with his contributors, the financial condition of the magazine was unvaryingly bad. Always hanging over his head was the danger that the *North American* would become hopelessly insolvent. He lived in terror that it might die disgracefully on his hands, or, as he expressed it in the phrase of a fastidious Brahmin, "go to some Jew." To cap his difficulties, his over-strained eyes gave him trouble for a season.[3]

But in spite of all these recurring exasperations he never relaxed his surveillance. Brief nervous jottings directed Lodge to solicit manuscripts, check proofs, stave off the printer, pay contributors, or run errands: "I go to Newport on Friday to stay over Sunday, so keep an eye on that last sheet and get us out punctually . . . You might ask Harry, or (better, if he would) Willy James to do Bayard Taylor's *Prophet* . . . If Weir Mitchell refuses Clarke, don't write to Dalton or anyone else till further notice. Damn C. C. Everett. Tell him the pressure for space etc., etc., we can't promise publication within any given time . . . Who is the best military critic in the United States? I want an article on Sherman's *Memoirs.*" Sometimes the printers got out of hand and needed to be sharply rebuked. "I wish the printers to understand that my orders are to be taken notice of and obeyed." Then to make sure, he himself addressed a sharp note to Mr. Clarke.

Like every thrifty editor Adams took especial pains to make manuscript copy and contract pages come out even. As a result he contributed more book reviews or "critical notices" than any one else. Of the one hundred and seventy-nine book notices that appeared in the eighteen issues edited by him, he wrote twenty-five or more. One of his chief aides was William Francis Allen, professor of languages and history at the University of Wisconsin, who produced fourteen notices. Marcel Thévenin, Répétiteur à l'École des Hautes Études, a "continental affiliation" picked up during his European tour, could be relied upon for expert reviews of learned German and French works on medieval history. As Adams used his magazine in part to direct scholarship toward Germany, Thévenin was called upon for a dozen notices. The assistant editor, Henry Cabot Lodge, obliged with eight reviews, as well as with an article on Alexander Hamilton. Among the more notable literary contributors were James Russell Lowell—three articles and four reviews; Henry James—one article and four reviews; William Dean Howells—one article; Francis Parkman—two articles. Palgrave's single contribution, "Thomas Watson the Poet," so delighted him— "a charming little article . . . just the kind of thing I most wanted"—that in long retrospect he considered Palgrave one of the two writers who kept the magazine literary, Lowell being the other one. His brother Charles was responsible for four articles and four reviews; his brother Brooks, for one article and two reviews. Even his father may have contributed two reviews for him.[4]

Fortified by his discussions with von Gneist, Stubbs, and Maine, Adams now offered (1873-74) the course in Medieval Institutions exclusively for "candidates for honors," and centered most of his interest upon this small group of talented young men, attempting for Harvard what Sir Henry Maine was accomplishing at Oxford in instituting the study of comparative early jurisprudence. Acting on a hint dropped by Professor Dwight in the Introduction to Maine's *Ancient Law*, Adams prescribed for Lodge "a course of special study on the early English law . . . with a view to ascertaining and fixing the share that the Germanic law had in forming the Common Law." With clear insight he had hit upon one of the major cruxes of English legal history. The larger possibilities of their joint labors becoming apparent, Adams proposed to President Eliot to establish a class of doctoral candidates to carry on the work "at his own expense." With the formal vote of thanks by the President and Fellows for "his generous action" to encourage him he went ahead to inaugurate graduate study in history at Harvard.[5] As soon as the new class could be established Adams disencumbered himself of one of his more elementary and more popular courses, "Political History of Europe from the Tenth to the Sixteenth Centuries," turning it over to Ernest Young, one of his doctoral candidates. He thus freed himself for a new course that he must have had in mind ever since the writer in *Old and New* had expressed the wish that the department might include "more of American history." In June 1874 he was "reading hard for a new course in colonial history" in which he expected "to expose British tyranny and cruelty with a degree of patriotic fervor which, I flatter myself, has rarely been equalled."

In 1875 the five-year term of his original appointment expiring, he accepted re-appointment for another five years at the same rank, assistant professor, and apparently at the same salary, $2000. Whether this arrangement reflected the financial stringency following the Panic of 1873 or simply Adams's indifference to matters of salary is not apparent. Perhaps his service was still too short to justify promotion to full professor, the next rank upward, particularly in view of the fact that his original appointment had been to rank unusual for a beginner. Of course, the most interesting aspect of the matter is that at the moment Adams was willing to commit himself for another five years. The occasion inspired characteristically violent reflections. "I am going on to thirty-eight years old, the yawning gulf of middle age. Another, the fifth, year of professordom is expiring this week. I am balder, duller, more pedantic, and more lazy than ever. I have lost my love of travel. My fits of wrath against the weaknesses and shortcomings of mankind are less violent than they were, though grumbling has become my favorite occupation. I have ceased to grow rapidly either in public esteem or in mental development." His correspondent, Gaskell, had no difficulty seeing through the pose. Outwardly, Adams may have begun to settle down; but discontent was in the blood.[6]

After two years work in colonial history he felt ready to

venture into the succeeding period in which his ancestors had played such crucial roles, the new course being History VI, "History of the United States from 1789-1840." How close the subject was to his heart we may infer from the terms of his offer to President Eliot: Henry Cabot Lodge was to inherit History V but substantially as Adams's deputy. "Mr. Lodge is to receive no pay from the university, and I presume will not be a member of the faculty. I am of course prepared to be responsible for his capacity." Whatever misgivings Eliot may have had about Lodge were apparently satisfied by this offer, for the catalogue for 1876-77 lists Lodge as an instructor in history. Of History VI, Lindsay Swift, one of the students who listened to Adams's lectures, has written: "Out of it grew, beyond a doubt, not only his largest work, *The History of the United States during the Administrations of Jefferson and Madison*, but also his admirable studies of John Randolph and Albert Gallatin, and his *Documents Relating to New England Federalism*." [7]

Essays in Anglo-Saxon Law

The hypothesis of historical development which made the greatest impression upon Henry Adams was the Teutonic Theory of history, the now largely discredited notion that constitutional and representative government had its origin in the folk life of the early Anglo-Saxons. In employing the hypothesis as a touchstone in all of his historical reviews, he aligned himself with the so-called Germanist school of Kemble, Freeman, Stubbs, and Green. But though he accepted the general theory of Teutonic origins, he relentlessly questioned the weaknesses of its supporting details. Stubbs, he contended, had been misled "by a sentiment of patriotism" to imply that Parliament was the outgrowth of the Anglo-Saxon Witan or royal council. Adams acutely observed that the old "witena-gemot perished with the class it represented" and that "two whole centuries elapsed between the last genuine meeting of the Witan and the first meeting of Parliament." Maine's "clever theories" that the English law of primogeniture was the product of Teutonic "tribal leadership in its decay" he regarded as equally unacceptable because the break in continuity lasted four centuries. Still, whatever the shortcomings of the theory it at least represented "a healthy reaction against the old tendency to consider everything good in civilization as due to Rome and Greece, to Cicero, to Homer, and to Justinian." [1]

Publicly, he clung to his authorities, German and English who assured him that the old theory was obsolete. Privately he felt that the new one still awaited scientific verification. To Sir Henry Maine he acknowledged that the "riddles of archaic law" grew more difficult the more he studied them. Yet he felt that without a unifying hypothesis there could be no history and no science of comparative jurisprudence at all. The alternatives were hypothetical unity or actual chaos. Unable himself to synthesize the enormously complex data of institutional history, he yielded to his passion for an all-embracing generalization and followed his masters in postulating unity. Uncertainty about fundamental premises was a vice of science that he could not endure. "I care little what theory is adopted," he once wrote to his friend Lewis Henry Morgan, the anthropologist. "One is for scientific hypotheses as good as the other. But I wish it might be settled which of the two shall be used as *the* scientific hypothesis." A similar unscientific haste affected his appraisal of Stubbs, whose scholarly caution irked him. On a "capital point" he believed that Stubbs should make up his mind: Either manorial jurisdiction was a continuation of hundred jurisdiction or it was not.[2]

At work in his mind were two powerful and contradictory impulses: on the one hand, he shared von Ranke's purpose to see the past "wie es eigentlich gewesen" and heroically excavated the documentary middens of German research. Because there "are long gaps of frightful ignorance" in our knowledge of the sources of English history from the fifth to the tenth century, he wrote in his review of Green's *Short History*, "the antiquarian must be here of more value than the generalizing historian." On the other hand, impatient to establish historical sequences he accepted almost as a tenet of belief Kemble's and Freeman's hypothesis that under the "purely democratic" early Anglo-Saxon constitution "the free citizens administered their own law and their own political system in their own courts." If the pre-Conquest legislative system was aristocratic and politically sterile, the same criticism, he was confident, could not be made of the judicial system.

The grand problem of his historical research became the investigation of early English history with a view toward forging the vital link that bound the remote democratic past to the democratic present. In this bold enterprise he enlisted the help of the most gifted students in his seminar in medieval institutions. The result of their collaboration, *Essays in Anglo-Saxon Law*, was studded with scholarship in the manner of the most formidable German models. Published in 1876, it included his own treatise, "Anglo-Saxon Courts of Law," and the dissertations of his first three doctoral candidates: Henry Cabot Lodge, Ernest Young and James Lawrence Laughlin.

"The Anglo-Saxon Courts of Law" stemmed directly from the positions maintained two years earlier in his daring review of Stubbs and supplied the demonstration of those venturesome pronouncements. For all his erudition, Adams had charged, Stubbs had no theory to advance that would make clear "the logical sequence of English political institutions"; consequently his narrative was "confused by details." He had not grasped the key significance of the relation of manorial jurisdiction to hundredal jurisdiction, which was that the civil as well as the criminal jurisdiction descended from the pre-

Conquest "hundred." If that link could not be established, English law would become a mere raft of institutions afloat in chaos. "The continuity of history," Adams said, "requires that this point should be strongly asserted." His present essay attempted that assertion.

The task which he set himself demanded that he prove first that the Anglo-Saxons brought not only the old substantive German law to England but also the courts of law and the German territorial district; second that the shift from the public jurisdiction of the hundred courts to the private or proprietary jurisdiction of the manorial courts, a process begun in the time of Edward the Confessor, placed the inception of English feudalism prior to the Norman conquest. The second conclusion he left to be confirmed in Lodge's able study of the development of the land law. In general, his thesis had been the doctrine of Palgrave, Freeman, and Stubbs, though Stubbs attached a somewhat greater importance to the Norman contribution to English feudalism. At a number of points, however, as a result of his fresh examination of the corpus of Anglo-Saxon laws, Adams placed himself in opposition to Freeman, and to Freeman's predecessor Kemble.

He was rigorously critical of English scholarship and equally rigorous in dealing with German writers. Rudolph Sohm might be impeccable but the great Schmid, he did not hesitate to say, reasoned falsely because he had ignored the habitual inaccuracy of the Anglo-Saxons in the language of their charters. By taking this inaccuracy into consideration Adams ingeniously traced the existence of several territorial hundreds back through a series of verbal disguises. He summarized the argument of the first part of the essay as emphatically as type would permit: *"The State of the seventh century became the Shire of the tenth, while the Shire of the seventh century became the Hundred of the tenth."* Out of this development came the three courts of the Anglo-Saxon judicial system: the witan (national assembly), the shire, and the hundred courts. Here was an acceptable historical sequence.

In the discussion of his second point, the rise of the manorial court as a private jurisdiction, which he dated from the relatively late time of Edward the Confessor and his ecclesiastical charters, he took issue impartially with the Englishmen, Kemble and Stubbs, and the German writer Konrad Maurer, all of whom argued for an earlier origin. In his opinion Edward's practices marked a revolutionary change in the Anglo-Saxon judicial system. "His acts show that he was not an Anglo-Saxon, but a Norman king. It was he who introduced the worst maxims of government into England." Granted that Anglo-Saxon legal institutions were marked by archaic inflexibility, the system before Edward had been redeemed from illiberality by its public and popular character. The manorial system of jurisdiction instituted by Edward, being essentially Norman in spirit, perpetuated the "most archaic and least fertile elements of both the Saxon and the feudal system." Through Edward's act "the theory of the constitution was irretrievably lost. Justice no longer was a public trust, but a private property . . . The entire judicial system of England was torn in pieces; and a new theory of society, known as feudalism, took its place."

What Adams intended to imply seems clear enough; the final establishment of public jurisdiction in modern times was not an innovation but a restoration of the liberal elements of the old Teutonic constitution, a happy recurrence to a vital first principle after the passing away of feudalism. The destruction of the manorial system and the patriarchal fallacy upon which it rested encouraged the return of better principles—democratic, popular, and free. Here again the continuity of institutions seemed certain.

Part of his demonstration turned on refuting Schmid's interpretation of "socn," in the medieval grants of "sac and soc," as granting "jurisdiction." By a piece of linguistic analysis that would have done honor to any hair-splitting German seminarist, he demonstrated that "socn" conferred fiscal jurisdiction, judicial jurisdiction being conferred by "sacu." It was an exquisitely learned tour de force and he was pardonably delighted with his own ingenuity. At the same time he had sufficient detachment to see the humor of his position. That he who had been schooled in the necessity of a nice adjustment of means to ends should have been seduced into such a parody of Dry-as-Dust in order to win reputation as a scholar was at least amusing. As some penance needed to be made if he was to assert his sense of proportion as a man of the world, he composed the following derisive epitaph in twelfth century Latin:

> *Hic Jacet*
> *Homunculus Scriptor*
> *Doctor Barbaricus*
> *Henricus Adams*
> *Adae Filius et Evae*
> *Primo Explicuit*
> *Socnam*[3]

In after years he made the tag a standing jest between him and his friends as if to prove that he knew that success in life demanded achievement of a quite different order.

Each of the essays in the volume corrected in detail various errors of the Germanist scholars. "My own position," he crowed to Gaskell, "will only bring your friend Freeman about my ears. I have contradicted every English author, high and low." He should have added that he and his students had equally taken to task the highest German authorities. None of the essays repudiated, however, the underlying romantic theory of Teutonic origins which was then at the height of its popularity, after the eclipse of French military power in 1870. According to Adams this theory gave "to the history of Germanic and espe-

cially English institutions a roundness and philosophic continuity which add greatly to their interest, and even to their
practical value. The student of history who attempts to trace
through two thousand years of dangers and vicissitudes the
slender thread of political and legal thought no longer loses
it from sight in the confusion of feudalism or in the wild
lawlessness of the Heptarchy, but follows it safely and firmly
back until it leads him out on the wide plains of northern
Germany, and attaches itself at last to the primitive popular
assembly, parliament, law court, and army in one which embraced every man rich or poor and in theory at least allowed
equal right to all." [4]

Henry Cabot Lodge in his study of Anglo-Saxon land law,
Ernest Young in his analysis of Anglo-Saxon family law, and
James Lawrence Laughlin in his investigation of Anglo-Saxon
legal procedure all followed Adams's lead. In each case the
chain of development was shown to be unbroken, however
much modified by the Norman Conquest. In each case the
writer identified the early Teutonic institutions with democracy
and individual freedom. Lodge argued that "on the nascent
feudalism of the Anglo-Saxons was superimposed the full-grown
system of William and Normandy"; but that before this happened there had already evolved the native ancestor of the
common law deed as evidence of title to land. Young documented Adams's destructive criticism of Sir Henry Maine's
Patriarchal Theory by showing that it was inapplicable to the
democratic organization of early German society. In that early
state woman was not degraded. The wholesome family organization of the Teutons had gradually degenerated in the historic
process which exalted "the individual at the expense of the
family as a whole." As Lodge categorically stated in his study:
"The fundamental principle of the equality of women before
the law, in everything relating to land, except the family land,
is indisputable." This last point having become peculiarly in-

teresting to Henry Adams, he reserved the full discussion of the place of woman in early society to a Lowell Institute lecture to be given later that year. Laughlin believed that there existed "a connection between English law and the primitive institutes of the early period of summary execution," and he took as one proof of early democracy the "larger judicial powers vested in the individual," his right to self-help under Teutonic law.

Dedicated to President Eliot of Harvard as the "fruit of his administration," the book also gave credit to Professor F. A. March of Lafayette College and Professor F. J. Child of Harvard College "for essential assistance in the translation of the Anglo-Saxon Charters" included in the appendix. For despite Adams's efforts to master the language the work turned out to be "the toughest bit of translation and editing" that he had ever attempted. That he did not rely wholly upon printed sources is suggested by the fact that in at least one instance he adopted "a few slight variations from Mr. Kemble's text," presumably as a result of his researches at the Bodleian in the summer of 1873. The cost of the publication, some $2000, he took entirely upon himself as the severe financial depression that still blighted the community had apparently not affected his own financial circumstances. Quite justly he expected to draw his "profit" from the success of his students for without undue vanity he could affirm that the book was "a really satisfactory piece of work." Not only was it a pioneer work of American scholarship; it could not be matched by any comparable achievement of the English universities.

His expectation of having Freeman about his ears went unfulfilled for his victim took no public notice of the book. As Freeman's intimate friend John Richard Green once reluctantly admitted, Freeman "with all his greatness profits very little by criticism." Stubbs, on the other hand, welcomed the book with a true scholar's urbanity by incorporating a number of its findings, with full credit given, in the following edition of his

Constitutional History. For a highly technical treatment of an obscure subject Adams's own essay was a model of lucid exposition and argument. Unfortunately it did not lack the little flourishes of triumph so dear to the German historical school. John Richard Green good-naturedly chaffed him on his German style, although he conceded that Adams was not quite so Teutonic as his pupils.[5]

The reviews were uniformly favorable. The *Law Magazine and Review* of London gladly hailed "so interesting and unmistakable a proof" that Americans "are awake to this fact of common inheritance." In Boston the *American Law Review* thought the essays "remarkable in the first place for their entire renunciation of English models and for their adoption not only of German methods, but of the authority and opinions of the now dominant German school." Approving Adams's intention, the reviewer added that "such monographs as these prepare the ground for a truly philosophical history of the law." William Francis Allen, who prepared a review for Adams's successor on the *North American,* took especial pleasure in the "masterly argument . . . upon the origin of manorial jurisdiction." The value of the book was seen by a fellow Germanist, Herbert B. Adams, at the just-founded Johns Hopkins University and he adopted it at once as a text in his historical seminar in Early Institutions.[6]

The eclipse of the Teutonic theory of democratic origins has, of course, invalidated the main assumptions of the work, but the light which the book casts on the survival of Anglo-Saxon laws still makes it a permanent and valuable contribution to the literature of the subject, according to recent bibliographies. As the historian Charles A. Beard has said "the persistence of Anglo-Saxon law" remains open to discussion "pro and con." Subsequent debate has shown that the sharp cleavage between the Romanist and the Teutonic schools is largely unreal. The assumption that early Teutonic society was "democratic" fitted in so conveniently to the ulterior purposes of the advocates of the Teutonic theory that they did not make a scientific study of it. Both sides, in their impatience to establish a "scientific" theory of descent, also glossed over the enormously complicated origin of modern political institutions.[7]

The strong partisanship for the Teutonic position was unquestionably inspired by the desire to bolster a political thesis. France, whose star had set once more at Sedan, conveniently provided the example of a nation that lurched wildly from a mongrel mobocracy to an equally mongrel despotism of a third Napoleon; whereas across the Rhine political reform proceeded not from the barricades but from the scholar's and jurist's study. It was the scholars and jurists who supplied the attractive rationale of the German Confederation, finding a noble ancestry for the political and military success of the New Germany in the early institutions of the German race. What was especially admirable, no doubt, to Englishmen like Freeman, Stubbs, and Green, and Americans like Henry Adams was that the German "revolution" did not seek to overturn the social and economic order. To the conservative statesmen who shrank from claiming even institutional kinship with the turbulent Franks, the racist theories of the German universities supplied new strength through ties of blood. It was the season, moreover, when the obligation of the modern Anglo-Saxon to civilize the world was widely honored by Anglo-Saxons in Germany, England, and America.

Henry Adams, brought up on Guizot, on Edmund Burke, and on the conservative philosophy of American Whiggism, which had been the faith of his father and his grandfather, had additional reasons for welcoming the Teutonic theory. In an age highly critical of the pretensions of democracy in America, the theory supplied an ancestry for American institutions that made them venerable and not merely virtuous. The ac-

count it gave of the origin of national sovereignty finally legit-
imated the scheme of government which his ancestors had so
largely helped to establish, as it traced it beyond the laws of
nature and of God to the primeval forests of Germany. The
defiant words of the preamble to the Constitution, "We the
People," were therefore not affirmations of rebellion but reaf-
firmations of the oldest political tradition of the Teutonic race.
Democracy was the true, underlying tradition; the divine right
of kings was, in Adams's words, "a mere historical blunder."
A government of checks and balances was also a grand bequest
of Anglo-Saxon and feudal times, for out of the conflict of interests between kings and nobles there arose "that partnership
of nobility and commonalty," again in Adams's words, "which
is the peculiar glory of the English constitution," [8] a glory that
shone all the brighter when transformed in the Massachusetts
constitution and the Federal constitution into a system of
checks and balances operating within a framework of separated
powers.

　These motives apparently prevented Adams not only from
questioning too closely the basic assumptions of the theory but
also from seeing the actual contradictions in his own argument.[9] On the one hand he believed that he and his colleagues
had shown the unbroken line of connection between modern
ideas of liberty and the ancient Teutonic ones; on the other
hand, when he discussed the actual mechanisms of survival he
acknowledged that "the manor was but a proprietary hundred,
and, as such has served for many centuries to perpetuate the
memory of the most archaic and least fertile elements of both
the Saxon and the Feudal systems." But if the liberal and fertile
elements of the Saxon system were not perpetuated through
the hundred system, by what mechanisms were they perpetuated at all? Conceivably the idea of political liberty—granting it had once existed—survived in Anglo-Saxon tradition and

legend. A mystical racist would affirm that it was a property of the Anglo-Saxon blood. Adams, venturing neither of these theories, silently ignored the dilemma which he had posed.

The Primitive Rights of Women

Not all of Adams's zeal as a censorious "teacher of teachers" was expended on his English colleagues. On occasion the ferule of the *North American* thwacked the knuckles of a compatriot scholar. In 1875 the flamboyant Hubert Howe Bancroft assembled the imposing volumes of his *Native Races of the Pacific* and nearly succeeded in cowing Eastern critics by the sheer magnitude of the project. Adams arranged for civil reviews of the volumes as they appeared, but soon sensed the gross shortcomings of the work, confirmed in his suspicions by the ethnologist Lewis Henry Morgan, whom he had solicited for a "ten or twelve" page notice of Volume II. Adams incited Morgan to make a full dress attack upon the volume, blandly advising him to ignore the favorable notice which the *North American* was about to publish. "I do not pretend to make the Review consistent," he reassured his accomplice, "except on moral questions." Morgan's merciless language delighted Adams, but he was even more pleased with Morgan's congenial thesis that the Aztec Confederacy was not monarchic, as Bancroft romanticized, but "essentially democratical." The desire of European historians to exalt monarchy had led to precisely the same kind of blunder that Adams believed he had discovered in English history. He greatly relished this unexpected confirmation of his opinions. "Our American ethnology," he assured Morgan, is destined to change the fashionable European theories of history to no small extent." [1]

Thus began a cordial correspondence between the two unconventional students of early institutions. After the success of Morgan's review, "Montezuma's Dinner," Adams immediately commissioned a second article, "The Houses of the Mound Builders," and at the same time began to bombard him with questions concerning the possibility of an Indo-European origin of social institutions. "I would like to ask you," ran his first query, "whether in your opinion the Aryan family, the Semitic, etc., had already before their separation, and before the occupation of America by our Indians, passed through any yet simpler stage of society or whether the stage we see in their common characteristics is simple enough to have served as the starting point." On another occasion, after a "warm argument" with Francis Parkman, who rejected the hypothesis of common origins on the ground that human nature was uniform in its development, he asked Morgan for more ammunition on the question "whether our American Indians had any trace of the political and judicial organization which characterizes the earliest Germans known to us."

The publication of Morgan's *Ancient Society* elicited the highest praise from his new ally; it was "the foundation of all future work in American historical science." It awakened all Adams's proselytizing zeal: "I have lost and shall lose no opportunity to impress on scientific men and institutions the need of careful scientific inquiry into the laws and usages of the village Indians." He assured Morgan that men like Professor Othniel Marsh, the Yale paleontologist, his own geologist friend Clarence King, Raphael Pumpelly of Harvard, Alexander Agassiz, and Major Powell would all give him "active sympathy." As for himself, the book gave fresh support for his belief in the German genius for "reconciling liberty with law." It also confirmed his opinion that the best materials for "the history of human progress" could be drawn from the action and reaction of legal conceptions.

One of the questions which he addressed to Morgan broached a subject which had increasingly come to interest him—the proper place of woman in civilized society. "Can you tell me,"

he began, "where I can find information on the customs regu-
lating marriage among our Indians?" At the moment his interest was severely practical for five months hence, December
9, 1876, he was to deliver in Boston his Lowell Institute lecture,
"The Primitive Rights of Women." The subject is of peculiar
importance because woman finally achieved an almost mystical
significance in his philosophy. Both of his novels—*Democracy*
in 1880 and *Esther* in 1884—were to present their heroines as
endowed with special moral and spiritual energies. And in
Mont-Saint-Michel and Chartres of 1904 he went on to analyze
the power of woman when it had achieved its highest form in
the "Queen of Heaven," who, in defiance of theology, reëstab-
lished the dignity of her sex. As for *The Education,* one need
only reread "Vis Inertiae" to learn that the special character
and destiny of woman have a central bearing upon the fate of
mankind. "Without understanding the movement of sex, history
seemed to him mere pedantry." Thus "as he grew older, he
found that Early Institutions lost their interest, but that Early
Women became a passion."

The Woman Question had affected the intellectual climate
of his earliest youth. When he was an undergraduate at Harvard the tracts of the Woman's Rights movement circulated on
the campus and were soberly reviewed in the *Harvard Magazine.* Later, in London, as a disciple of John Stuart Mill, who
was the foremost English champion of the rights of women, he
could hardly escape taking sides on that vital issue. Then in
1869 came Mill's great essay "The Subjection of Women,"
arousing its storm of comment wherever thoughtful readers
forgathered. Issue could now be fairly joined on the highest
level of discussion. Of course, in the more sordid realm of political action, Henry Adams doubtless shared James Russell
Lowell's amused contempt for the perspiring crusaders who
proclaimed the New Jerusalem in strident treble voices. Entirely beneath serious notice were the unseemly activities

of Susan B. Anthony, Elizabeth Cady Stanton, and Lucretia Mott, and their National Woman Suffrage Association. Henry's eldest brother, John Quincy Adams, stated what must have been the family position on the subject when, as candidate for governor of Massachusetts in 1871, he wrote that he was opposed to woman's suffrage because there was between the sexes a "division of activities and function which seem to lie at the foundations of society." Woman's suffrage did not touch the really basic question: How to establish the dignity of women as co-equals with men? In his judgment mere legislative enactments did not face that question at all.[2]

Every chivalrous impulse impelled Henry Adams to revolt against the legal, as distinguished from the political handicaps, imposed on woman; for the legal barriers implied inferiority. His lecture might well have been entitled "A Vindication of the Legal Rights of Women." He did not need Mill's essay to prove to him that the degradation of women grew out of convenient myth and not out of the nature of woman. As a descendant of Abigail Adams he knew woman's capacity for greatness within her sphere. He had a similar respect for his mother; in one of his early letters from Germany, he confessed how "like a selfish, low-minded fool" he felt after certain long talks with her. Moreover, within the family circle had sparkled the bright intellect of his elder sister Louisa whose death in 1870 had left him with an oppressive sense of wasted talent. In London he had also learned the importance of women in the great world of public affairs, where their brilliance and beauty had captivated him. And now in his own household the flashing wit and self-assurance of his wife made the myth of inferiority even more incredible.

Taking a hint from a lecture given in London by his friend Sir Henry Maine in 1873 on "The Early History of the Property of Married Women," Henry Adams ignored the contemporary hue and cry for new laws and devoted his lecture to stating the feminist position of the enlightened conservative. In a sense it was an extended rebuttal of Maine's position, as first stated in his *Ancient Law* and reaffirmed in his lecture of 1873. Adams's main purpose was to disprove "the generally accepted opinion of writers on Primitive Institutions" that "the original position of married woman was one of slavery, or akin to slavery," and that the history of woman was her gradual rise and escape from this state of original degradation to her "complete triumph" under the influence of Christianity. Sir Henry Maine's theory that in the earliest society the "husband acquired over his wife the same despotic power which the father had over his children," seemed to him inadequate to explain the history of woman's status and rights.[3]

He conceded that "probably the institution of marriage had its origin in love of property," which was one of the strongest and most energetic of human instincts.[4] Granted this fact the question was, whether in primitive societies wives were deemed ordinary property. Lewis Henry Morgan had shown that among the early American Indians the wife remained "a free member of her clan," however shifting the marriage relation might be. Since, said Adams, the American Indian was a branch of a primitive Asiatic stock and had been little affected in his development from without, the "free" state of his womenfolk implied a similarly free original state in primitive Asiatic society. Egyptian antiquities, as described by Mariette Bey, indicated a parallel development, proving that the social position of woman was "highest in the ages most distant." In the Egyptian Trinity, for example, the woman, "Isis, stood on the same plane as Osiris." When the Christian theologians adopted the Trinity they "dethroned the woman from her place," but the "irresistible spread of Mariolatry, the worship of the Virgin Mother, proved how strongly human nature revolted against" this degradation of woman.

The Aryan races—Celt, Roman, Greek, German, and Scandinavian, in their primitive stages showed a similar freedom from degradation. A most dramatic illustration in Greek history was the story of Penelope and her suitors. As for the Romans, though they subjected women to a severe legal code, that code was opposed to original Roman family institutions. In time, by various legal expedients, the wife was emancipated from the husband's authority until by an excess of reaction "the whole family organization was shaken to pieces." But even by the terms of that severe code the wife was not really degraded; the law did not discriminate against her because she was a woman but because she was by a fiction regarded as a daughter and hence subject to the system of filial law and custom that bound all. As for the Teutons, the *Njalsaga,* the Icelandic equivalent of the *Odyssey,* exhibits a heroine no more a slave than Penelope. "All the fierce and untamable instincts of infinite generations of fierce wild animals were embodied in her."

If the history of institutions showed that woman was not strictly property, nor a slave in the early Christian era, how could one account for the elements of degradation that survive in Continental law and especially in the English common law? The answer was that the Church was the principal agency in degrading the status of women. "It rose to power under the intense moral reaction against the corruptions of the Empire; and of all the corruptions of the Empire none had been more scandalous than the corruption of the women!" Steadily the ascetic principle in religion worked its alterations. In the Church economy, duty and submission were more important than civil rights. And as the status of woman declined she became more and more dependent upon the Church as the chief source of protection and consolation. In time the "pure humanitarian influences of Christianity" operated to relieve the "manifest injustice toward women which the Church had either stimulated or permitted." By that time the old freedom of the Roman law was gone and the barbarian independence of the

Teutonic wife could not be restored. "The Church established a new ideal of feminine character, thenceforward not the proud, self-confident, vindictive woman of German tradition received the admiration and commanded the service of law and society . . . The Church raised up, with the willing cooperation of men, the modern type of Griselda,—the meek and patient, the silent and tender sufferer, the pale reflection of the Mater Dolorosa."

But it was more than the moral reaction that produced this transformation. In the social and political chaos of medieval times "the most pressing necessity of society was concentration." The Church seeking to create "a new unity and a new morality for mankind" required a paramount authority. Unable to rest it upon the will of a disorganized society, it resorted to "the divine will." For analogous reasons the arbitrary monarch, the feudal noble, and the husband of that day were precluded from resting their respective claims to authority upon the consent of society. "Thus the family, like the state, took on the character of a petty absolutism; and to justify in theory the sacrifice of rights thus surrendered by the wife and children . . . men fell back on what they called the patriarchal theory, and derived the principles they required from a curious conglomeration of Old Testament history and pure hypothesis."

Woman's civil rights had thus been impaired by precisely the same fiction, by the historical blunder later adopted by Hobbes and Filmer, which obscured the originally free character of the English people. Cromwell and Luther were therefore conservatives in a deeper sense than they imagined, for in recurring to earlier free institutions they "represented more than the protest against religious or political absolutism." They were instrumental in producing "a readjustment which the exigencies of a pressing immediate necessity had for a thousand years thrown from its natural equilibrium." Central to that equilibrium was the constitution of the family, because the social equilibrium depended upon the equilibrium of the family. The family in turn drew its strength from man's two chief natural instincts: his affections and his love of property. Either the excessive development of authority or the excessive decline in family authority, each being a disturbance of that natural equilibrium, must produce its reaction. Whatever the reaction, the family would survive because it is "the strongest and healthiest of all human fabrics." [5]

If there were any representatives of the New Woman in his audience, they could hardly have relished this learned disquisition on the Woman's Rights question. True, it gave the social and legal rights of women as noble an ancestry as the political liberties of men but its strong emphasis upon the family as paramount to the individual played into the hands of the pulpit demagogues who opposed any extension of woman's rights on the ground that they threatened the purity of the American home.

On the side of scientific history the lecture offered Adams another opportunity to study the causes of historical phenomena and to deduce the laws of their development. Like his friend Lewis Henry Morgan, he threw caution aside once he felt himself on the track of an all-inclusive generalization. Later scholars have pricked the iridescent bubble of these "hasty generalizations," seriously questioning the hypothesis adopted by Henry Adams of "universal stages of evolution through which all mankind has run." Regardless of the validity of its main premises, however, the essay is an astonishing tour de force, daringly synthesizing a great mass of historical, anthropological, and legal data. Of all of Adams's early writings it is the most sustained and original piece of historical reconstruction. [6]

It happened that on the following evening Susan B. Anthony gave an address advocating women's suffrage in which she declared that only through the ballot could her sex hope to redress its immemorial grievances. With masculine astuteness the editor of the Boston *Transcript* printed a summary of Adams's lecture immediately below that of Susan Anthony's so that the final sentence seems a direct admonition to the hopeful suffragist. "In conclusion," wrote Adams's reviewer, "he repeated the scientific statement that every new civilization is a reproduction of an old one, and said that if women succeed in this age or any age to come, in attaining complete and entire independence, they will merely be reviving the same principle as was once a part of the social scheme of Rome." [7]

American History and the Constitution

In his more optimistic moments Adams believed that the study of early comparative jurisprudence might train American lawyers to be historical jurists. In moments of doubt he felt that the rigorous study taught at least a method of investigation and a discipline of the mind as good as mathematics. But whatever its value to his students, he cherished no illusions about its value to himself. The impulse toward practical ends had led him into the half-world of conjecture and hypothesis where he was oppressed by the endless twilight and haunted by endlessly receding vistas of *sac* and *soc*. For himself, he had mastered method. It was time to approach the nearer objects of study, which, as he pointed out to Lodge, were the historical forces at work in modern England and America. His ancestors had been personally absorbed in establishing a point of equilibrium. He could make it his mission to evaluate their success.

The course in the history of medieval Europe having yielded its juice, he discarded it. Medieval institutions—chiefly Anglo-Saxon—provided an anchor for American institutions in the remote past of England and Germany. Thus prepared he moved into the next field, the legal and constitutional history of England to the 17th century. This study brought him to the threshold of American colonial history, and he began the investigation with his students of the long foreground to 1789. The final phase of preparatory study quickly followed: History VI, "the history of the United States from 1789 to 1840." The long thread of human liberty, which he believed he had traced back to its beginnings in the forests of Germany, brought him out at last to the great westward movement of democracy into the forests and prairie lands of Trans-Mississippi. As democracy and constitutional liberty had been the central theme of his medieval studies, it was equally central to the study of these later periods and made the whole vast expanse of historical development cohere.

The reviews which he wrote for the *North American* discussing current works in American history provide a cross section of the historical principles which he adopted in the course of his new researches and they indicate what was to be the main subject matter of his future work in American history. A forecast of his approach appeared in the criticism of Bancroft's interminable history, which had just reached, in 1874, its tenth volume. The review, to which Adams contributed the section on diplomatic history, spoke rather scornfully of Bancroft's naïve faith in the abstract virtues of democracy and the "gentle feelings of humanity." Man might indeed be perfectible but he was innately selfish and the rise of Jacksonian democracy showed the measureless depths of that selfishness. Hence for all his great and admirable knowledge of source materials, Bancroft lacked the "strictly judicial mind," the review said, and his strong partiality for America occasionally led him into misstatements of fact. His most striking success lay in general in the diplomatic history, but Adams quarreled with Bancroft's treatment of the 1783 peace negotiations with England because it diminished "the natural effect of the drama." By raising the character of Lord Shelburne, the English negotiator, Bancroft lowered the character of Fox, who, with Burke, had proposed a more manly course of action, direct recognition of American independence. As an English statesman, said Adams, Shelburne was in fact weak and devious; and though America profited from his weakness the world could "hardly be expected to admire his abilities as a statesman." Artistically the drama required that Franklin's triumph over his adversaries should be shown to be absolute. In this judgment Henry differed from his father who had denied Fox "a place among Britain's best or purest statesmen," largely because of his refusal to coöperate with Lord Shelburne. This place Henry restored to him. Only obliquely did Henry allude to John Adams's part in the peace negotiation and then mainly to praise Franklin's consummate skill in guiding his colleagues, especially the choleric John Adams, who, according to Franklin, was "sometimes and in some things absolutely out of his senses." [1]

In Colonial and American history John Gorham Palfrey was his chief adviser, and Adams relied on him for authoritative reviews of books on New England subjects. Palfrey generously accommodated him although he was in the midst of his labors on the fourth volume of his *New England History*. When the volume appeared in 1875 Adams praised it extravagantly. Age and illness, he averred, had not impaired Palfrey's work. "The qualities which in our opinion have hitherto placed Dr. Palfrey absolutely first in the ranks of American historians, the strong good-sense, the thorough study, the sober and finished style, the contempt for sentimentalism and affectation either of thought or manner, the lurking humor, and, above all, the thoroughly healthy and manly insight into the morals of the subject, seem as evident in this volume as in any that have gone before." Palfrey's great work showed how far the modern age of "diluted morality and popular history" had fallen from "that strong quality of mind and thought which was so characteristic of the Puritan age," in its greatest epoch. That period which ended in 1689 had been a time of large and liberal thought; religion was still a living force; and men's tempers were amiable. The breakdown of the religious commonwealth ushered in a period of decay, paralleling a similar decay visible throughout the Western World. In England both the church and the constitution were seriously threatened, but fortunately the continuity of their development was preserved by Whitefield and Chatham. In France and Germany, on the other hand, the convulsion "destroyed the continuity of their history." Adams regretted that Palfrey had not supplied the financial history of the period because the decay had infected financial morals as well and brought on a period of speculation and currency delusions. In his opinion, "all the most radical financial theories of 1875 were put into full practice a century and a half ago in New England." [2]

What Adams regarded as especially significant in New England was the long and "successful constitutional struggle against the influence of the Crown," a struggle which established "precedents of which no one else in the whole world then understood the value." By using the town meeting as a democratic instrument of resistance, the descendants of the Puritans learned to act as one and obliged history to deal with them thereafter only "in mass, as a tendency, a force, which belonged to the soil and the atmosphere." That development signalized the decline of the individual. In that respect it marked loss, however inevitable and necessary the change, and foreshadowed the future descent of American society toward democratic mediocrity.

In the article which he jointly wrote with Lodge in 1876 reviewing von Holst's *Constitutional and Political History of the United States*, Adams disclosed a drift away from the a priori conceptions which had characterized his Civil War view of the Constitution. Earlier, when he had been preoccupied with the effect upon the Constitution of the struggle over slavery, he had reasoned about it in terms of purely legal theory and idealistic preconceptions. It had, so to speak, an independent and inviolable character, but after Grant's inauguration he came to believe that "essential and fatal changes" had affected that organic law, and that these were the result not of the Civil War "but of deeper social causes." Von Holst helped him to see that his earlier analysis had been too doctrinaire, too perfectionist in its approach. And he saw this precisely because von Holst himself had set up "an absolute political standard so high" as to be practically unapproachable, a standard that made America seem a "lamentable failure." Adams was now ready to concede that the Constitution had not met the "theoretical requirements of the political situation" any more than the great documents of the English constitutional system; but he contended that the American Constitution deserved to be venerated because it had succeeded so remarkably from the practical point of view.

The problem had been "to form a more perfect union" without resort to force; that was the overriding necessity. The chief obstacle which had to be overcome was the strong tendency toward "particularism," the states' rights heritage of Colonial days. That conflict has taught us, said Adams, that the doctrines of nullification and secession which plagued pre-Civil War America "were not the creatures of the slavery question" but were as old as the Union itself. It seemed to him that the "only real question" was "how the Constitution has performed its work of union in the face of these particularist tendencies." Its success deserved therefore to be measured by the effectiveness with which it implemented the movement toward nationalism. The Constitution so perfectly adapted itself to the irrepressible movement of American institutions toward nationalism, he pointed out, that the party in power was always obliged to adopt the nationalist point of view and the party out of power was driven to assert the states' rights position. This principle he made one of the maxims of his teaching. "That Jefferson

One of von Holst's most provocative sections was his treatment of the War of 1812, for though he acknowledged that the
war did much to create an American nationality, he deplored
as discreditable the intrigues which brought it on. Adams defended the whole transaction as having restored American self-
respect and strengthened the national feeling more than all
the twenty-five years preceding it. One of its best results was
that "New England learned then, once for all, not to trifle with
the Constitution and with the Union." Moreover, Jefferson's
administration was the first to show the inevitable operation of
the law of national development by which "the precedents
established by the Federalist administrations were accepted
and enlarged by the Republican administrations." Obviously
generalizations like these would require extended proof. To
carry conviction, the historian would have to show the operation of the law of centralization in all avenues of American life.
It would require him to measure the movement of the social,
economic, political, and intellectual forces of the United States
during the period when the national spirit first emerged. The
review suggests that such an enterprise was already germinat-
ing in his mind and within a few years would flower as the
plan of his monumental *History of the United States during the
Administrations of Jefferson and Madison,* a work which should
scientifically refute foreign criticisms like those of von Holst.

The German critique of America probably had a more immediate effect upon Adams's study for his new course, the "History of the United States from 1789 to 1840," and may, in fact,
have determined its emphasis. One of the most disputed points
of the history of that period was whether the New England
Federalists had planned or attempted treason against the federal government. Von Holst in his account of the Campaign
of 1828 told how the partisans of Andrew Jackson, aiming to
discredit President John Quincy Adams, had revived the charge
that he had been guilty of a "betrayal" of his own party in
1807 when as a Federalist senator from Massachusetts he sup-

ported the Embargo Bill of President Jefferson, against the advice of all of his party associates. To defend himself, President Adams had countered with the sensational accusation that the leading members of his own party had not deserved respect because they had dishonored themselves at the time in the separatist conspiracy of the Essex Junto. The Boston Federalists in an "Appeal to the People" vigorously denied the accusation, but President Adams failed unaccountably to present his proof. Von Holst concluded that "the final decision of history must therefore be suspended" with respect to Adam's veracity.

The question involved not only Henry Adams's ancestor but also George Cabot, the ancestor of his assistant Henry Cabot Lodge, who had been the leader of the Essex Junto and the president of the Hartford Convention. Von Holst's dramatic challenge came at a fortunate moment in the work of Adams and Lodge. Several months before the two men had begun to read von Holst, Lodge had begun digging into the eighty-eight volumes of the Pickering Papers for a projected life of George Cabot. Lodge kept Adams posted on the progress of his excavations. They soon became aware that Charles Upham's recent life of Timothy Pickering, who had been one of the leaders of the Essex Junto, gave a completely erroneous impression because, as they discovered, incriminating papers had been suppressed. Henry Adams unearthed even more interesting evidence in the Adams family archives where for almost a half century had reposed the document which might have resolved von Holst's doubts. It was John Quincy Adams's devastating reply to the Federalists' "Appeal to the People." With this remarkable polemic, that some motive of prudence had caused to be suppressed, and the series of incriminating letters from the Pickering Papers in the Massachusetts Society Collections, Henry Adams was in a position to prove the uncompromising patriotism and integrity of his testy ancestor before the bar of history.

Lodge published his *Life and Letters of George Cabot* by

the middle of the following year, 1877. Adams reviewed it for the *Nation* and in a fashion that he told Lodge was "ingeniously calculated to make everyone, yourself included, furious with indignation." Though he praised the scholarly method of the book, he carefully pointed out that it was "avowedly partisan," Lodge's aim having been to present "his ancestor in the most favorable light." Lodge had of course made every effort to dissociate Cabot from the extremist faction headed by Pickering and had put him in the more respectable company of Alexander Hamilton and Rufus King. Adams then moved to present the evidence in support of his grandfather's probity. This appeared in the same year as *Documents Relating to New England Federalism, 1800-1815* with a preface which asserted that "So far as the editor is concerned, his object has been, not to join in an argument, but to stimulate, if possible, a new generation in our universities and elsewhere, by giving them a new interest in their work and new materials to digest." The preface in alluding to these "Northern schemes of resistance" against the federal government seems almost to offer the work as an act of expiation, for New England had got off scot-free despite her sins; whereas the South had been crushed and brutally reconstructed.

The collection of documents was the first published fruits of Adams's studies in American history. Godkin showed his senses of poetic justice by printing in the *Nation* a review of the *Documents* by Henry Cabot Lodge.[4] Lodge professed to see in the collection confirmation of his view that there were two elements among the Federalists, the extremists and the moderates, the latter, including Cabot, being opposed to desperate measures. Returning tit for tat he called attention to the suppression by the editor of "the most personal passage" of invective relating to Otis contained in John Quincy Adams's "Reply." The omission, he said, was a tribute to Adams's chivalry if not to his scholarship. As Lodge omitted any other accusation of family bias, Adams good-naturedly wrote, "I congratulate you on the judicial elevation you are attaining."

There can be little question that Adams's prepossessions were toward political liberalism. "There was a legend in college," Lindsay Swift has written, "that Mr. Adams conducted the democratic side of American history and Mr. Lodge the Federalist or 'aristocratic side.'" When Lodge prepared his essay on Alexander Hamilton for the July 1876 *North American* and praised Hamilton's administrative expedients, Adams passed the article, after making many verbal corrections, but strongly dissented from its thesis. "I dislike Hamilton," he explained, "because I always feel the adventurer in him. The very cause of your admiration is the cause of my distrust; he was equally ready to support a system he utterly disbelieved in as one that he liked." His "innate theory of life" was wrong. Their own ancestors, he reminded Lodge, had fortunately left as a legacy a very different "moral standard." He saw no substantial reason, he said, for not holding to his inherited feelings "toward Jefferson, Pickering, Jackson and the legion of other life-long enemies whom my contentious precursors made." In the case of Hamilton his feelings went beyond the traditional ones and constituted his "own kind of aversion." Still as a scholar, he believed that the rival point of view should be heard. Hence when he proposed to Eliot that Lodge should conduct a rival course in history, he said: "His views being Federalist and conservative, have as good a right to expression in the college as mine, which tend to democracy and radicalism." [5]

Political Reformer: Final Phase

Fruitful as his work was in his role as scholar and teacher it confined his speculations within academic limits. The editorial chair on the other hand was an incitement to speculations of the widest sort concerning contemporary American life. The approach of the Centennial year suggested to the editor that a scientific survey of the progress of American civilization might appropriately be made for the purpose of predicting its future lines of development. He recruited six experts in as many principal fields of American thought: Simon Newcomb in science, J. L. Diman in religion, William Graham Sumner in politics, C. F. Dunbar in economics, G. T. Bispham, a Philadelphia lawyer, in law, and Daniel Colt Gilman, president of Johns Hopkins University, in education. His plan, he explained to Gilman, was "to measure the progress of our country by the only standard which I know of, worth applying to mankind, its thought." The mode of treatment was not prescribed, but the common formula may be gathered from the suggestions given to Gilman for his conspectus of American education: "The American conception of Education, therefore, as shown in her efforts during the last century; the changes in that conception, if any; the strong and weak points of her ideal and its realization; and the degree of influence which may be justly claimed for it, on Europe; finally the prospect for the future; these views combined make a subject of the highest interest and permanent value." [1]

Such an investigation would round out the pattern of development of the national character which Adams was already studying in its constitutional and political aspects. Taken together these inquiries should disclose the laws of development of the American mind. What he appears to have had in view was well expressed in a quotation from the historian Taine whose philosophy had been the subject of a full length article in the *North American* for October, 1873. "The historian," said Taine, "notes and traces the total transformation presented by a particular human molecule or group of human molecules, and, to explain these transformations, writes the psychology of the molecule or group . . . The task is invariably the description of a human mind or of the characteristics common to a group of human minds." To achieve that task of scientific description the historian needed to adopt, according to Adams, 'some basis of faith in general principles, some theory of the progress of civilization which is outside and above all temporary questions of policy." [2]

His "Centennial" issue did not depict, however, a uniformly progressive motion. In fact, upon the thoughtful reader the effect must have been decidedly sobering. The possibilities of progress were recognized, but formidable obstacles stood in the way. Pure science languished while technology and the practical arts flourished. American politics were cankered by widespread corruptions. In political economy we had yet made no progress toward investigating the "laws of material wealth" because we were still able to squander our natural resources. In religion there had been a great increase in practical morality but a precipitous decline in theology. In education an immense diffusion of knowledge had gone on at the lower levels of instruction, but above the primary grades the system lagged. Only in law could the scientific observer assert unqualified progress; the law had become "more simple, more humane, and more adaptive." The survey of American law was especially notable for its acceptance of Buckle's dictum that "the *causes* of civilization are to be found either in external influences or in internal mental action." It followed that the character of national development is controlled by physical causes—political and historical—social causes, and intellectual causes.[3]

Godkin, of the *Nation,* who conscientiously boosted the *North American* at every opportunity, published a long review praising the remarkable Centennial issue. He regretted, however, that the tone of it might limit its usefulness. "The 'practical man'—or, in other words,—the Philistine—comes in for some pretty hard knocks, a fact which will probably prevent the number from having a wide circulation among members of Congress or among the aldermen and mayors of large cities.

This is a pity, for what the Philistine needs to know is the depths of his own ignorance." [4]

If Adams's "Centennial oration" in the von Holst review voiced his greatest hopes for the future of America, the Centennial issue voiced his deepest fears. These fears kept him firmly allied with the reform element in politics no matter how great the pressure of other affairs. During his year's stay in Europe the interim editor of the *North American,* Thomas S. Perry, had largely avoided reform politics. On his return Adams determined to reopen the pages of the magazine to the reform group. As before, he turned first to Jacob Dolson Cox, whom he had long thought best equipped to take the intellectual leadership. Then he wrote to his friend James Abram Garfield to explain his predicament: "I *must* have a man to write me a political article for the *North American* for April [1874]. I wrote to Cox, but of course he can't. I suppose you won't and it is no use to ask you. But do you know any man in Congress or in Washington who has ability and knowledge enough to do it? I want a horoscope cast; a birds-eye view of the situation; a vigorous statement to our friends of what they can hope to do and what they can't, and whether their wisest policy is to organize inside the party or out of it. Tell me whom I could apply to. Schurz himself is the right man, but I fear he cannot do it." [5]

Preoccupied with his own struggles in Congress, Garfield gestured in a friendly fashion and declined. As a result reform went unserved in the April number. Then on March 12, 1874, Charles Sumner, senior senator from Massachusetts and spiritual head of the Republican party of Massachusetts, died. An unseemly clamor arose over the succession. Sam Bowles of the Springfield *Republican,* long a partisan of the Adamses, urged the selection of Charles Francis Adams. Henry's father, then in retirement was, against his wishes and uselessly as it turned out, stirred from his long repose. Henry could pretend to "laugh at the whole concern," unhappily secure in the knowledge that his side was "commonly beaten in politics," yet he confessed that the incompetence of the government was worse than corruption and "enough to make one a howling dervish for life." Such pleasantries might do for his English friends. However, when Sumner's mantle was finally draped about George S. Boutwell, the nonentity who as Secretary of the Treasury had epitomized administrative incompetence in Adams's eyes, he was driven to action. If ranking officers could not be found to direct the campaign, subalterns must serve. Adams enlisted his outspoken younger brother, Brooks, and the always dependable Charles.

After its long political holiday, the *North American* marshalled its phalanx of writers. Brooks Adams, admitted to the Massachusetts Bar only a year before, struck off an article for the July 1874 issue, "The Platform of the New Party," whose authoritative style might well have passed for Henry's own. In a measure it was a reworking of Henry's scathing "Civil Service Reform." Like that appeal it invoked the dogma of the Massachusetts Bill of Rights that "a frequent recurrence to the fundamental principles of the Constitution . . . is absolutely necessary to preserve the advantages of liberty and to maintain free government." In Grant's administration the spoils system had finally captured all branches of the government and had thus broken down the system of checks and balances inherent in the Constitution. The party leaders in caucus now constituted an invisible and rival government which controlled all branches of the regular government. If no remedy were effected the nation might at last be driven to choose between "anarchy and disintegration, or force." Brooks's statement was supplemented in the same issue by Charles's long article on the "Currency Debate of 1873-74," which forcefully asserted the conservative position against inflation of the currency. One by one he scornfully pilloried the inflationists, but his best strokes he reserved for Messrs. Butler and Boutwell.

Even though that July issue may have had the air of a family publication, Henry Adams could at least be satisfied that he had put the *North American* in the van of the reform movement. Henceforward in almost every issue he harried the Philistines. In October David A. Wasson delivered a weighty homily, "The Modern Type of Oppression," on a theme touched by all of the Adamses since the close of the Civil War: the concentration of irresponsible and corrupt power in private hands. Gamaliel Bradford in "Lombard and Wall Streets" drew upon England's financial experiences, as Adams had done ten years earlier, to suggest the proper procedure for moving toward the resumption of specie payments. On the political front, Charles F. Wingate, working in collaboration with Henry's elder brother Charles, began his exposé of the Tweed Ring under the title "An Episode in Municipal Government," the first of a series of four articles.

For the rest of his term as editor Adams could count on aid from only two of his old Washington colleagues; Francis A. Walker helped him out with a carefully documented criticism of the classical wage fund theory and gave academic comfort to the labor reformers who sought higher wages. Of more practical bearing was David A. Wells's article on the "Reform of Local Taxation," which he wrote only after long urging. Adams had reminded him of the "evils and absurdities" of the tax system, and as a starting point he suggested that the Massachusetts tax system should be "ripped up without mercy." [6] Wells followed the directive with considerable success by exposing the confusion and deceptions always produced by efforts to levy personal property taxes.

Articles in the *North American* might help the good cause, but much more would be needed to defeat the continuation of Grantism at the next election. To do his part Adams needed first to reconnoiter Washington. He had not lost contact with his friends of the capital. At every opportunity during college recesses he hurried down to Washington to draw strength like Antaeus from the raw earth of politics. In the winter of 1875 on one of those life-giving visits he set about to "organise a party of the centre" which might wield the balance of power and impose a reform candidate upon one of the major parties. It was a daring scheme, requiring strict secrecy and "devious and underground ways," pleasurably exciting to one bored with Boston.[7] One of the first requirements was a newspaper to serve as the organ of the group. He briskly entered upon negotiations to acquire the Boston *Advertiser* and commanded his new political lieutenant and graduate student, Lodge, to join the syndicate. For some reason the plan fell into abeyance; but the work of organization flourished. The little nucleus formed in Washington busied itself under Adams's energetic direction to promote a meeting of notables in New York.

In April 1875 he reported to his chief, Carl Schurz, that "Sam Bowles, my brother Charles, Henry Cabot Lodge and I concocted a letter and issued it . . . to such gentlemen who were on your lists." He advised Schurz to prepare an acceptance of the invitation that was to come from the New York committee. He also kept in close touch with David A. Wells. Some sort of working organization was effected at the April meeting at Delmonico's and Adams returned to Boston filled with new hope. By common consent Lodge had been made secretary of the Boston correspondence committee. Adams, buried under his teaching and editing responsibilities and hoping for vicarious triumphs, urged his young friend to make the most of his opportunities: "Anything which takes a man morally out of Beacon St., Nahant, and Beverly Farms, Harvard College and the Boston Press, must be in itself a good." [8] Success would put him "in a very commanding position." Even in the case of failure he would have won "a wide range of acquaintance and influence in and out of this mouldy little community."

Hayes's victory in the Ohio campaign under the leadership

of Carl Schurz invigorated the Independents, for the keynote of the campaign had been currency reform, i.e., resumption of specie payments. Quite correctly Adams realized that that victory might enable his friends to control the election. He did not foresee that he would be unable to control his friends. The first scheme of the new organization had been to force his father's nomination upon the parties; this had been Schurz's wish. The elder Adams, who had already seen so many political bubbles burst, showed no enthusiasm for the project. Like his forebears he would accept public office, but he would not seek it. Even his strongest supporters could not argue that he would be a truly popular candidate. Hence the scheme had to be abandoned, much to Henry Adams's relief. As he told Schurz, "I am not sorry for it. I do not like *coups de main*. I have no taste for political or any other kind of betting, and for us to attempt forcing one of ourselves on a party convention, necessarily entails jockeying of somebody." And jockeying would do permanent harm since by making use of the caucus system they would strengthen it. He believed that they should center their attack on the caucus system, "the rottenest, most odious and most vulnerable part of our body politic."

He suggested in great detail to Schurz the strategy that should be adopted. Lacking sufficient public support to set up their own man they must turn to the one who came nearest their standard, Benjamin H. Bristow, the Secretary of the Treasury, who had won fame for his prosecution of the "Whiskey Ring." Of course Bristow was a strong party man whom the Independents could support only on grounds of policy in the expectation that as president he would "in time work round to our opinions." Bristow would not accept an Independent nomination; hence the Independents must work with Bristow's friends, but avoid declaring for him until the Cincinnati Convention met. If Bristow should receive the regular nomination, that was the most that could be hoped for in the present campaign. The cardinal need was to act as a unit and to agree upon "resistance to caucus dictation."

He proposed that Schurz should call a meeting of some "two hundred of the most weighty and reliable of our friends" after the close of the Convention to decide on a course of action. To speed the project Adams submitted a draft of a suitable circular letter. Schurz countered with the suggestion of a pre-Convention conference endorsing Bristow, who had just resigned from Grant's cabinet. The draft letter though "excellent" would need to be amended "to serve a different purpose." To soften his disagreement, Schurz assured Adams, "I shall be very grateful to you if you will let me have the benefit of your advice as often as possible." [9]

Meanwhile Lodge had run down to Washington, largely as Adams's emissary, and was immediately "plunged up to the ears in Washington intrigue." Adams cautioned that if it was necessary to telegraph, "Schurz and Bristow had better figure as Smith and Brown." The newspaper scheme reappeared, this time looking toward the purchase of the Boston *Post*. In Lodge's absence in Washington the chief work of negotiation, including organizing a purchasing syndicate, fell upon Adams. He planned to put in $5000 himself toward the $150,000 needed. While urging Lodge to get "Bristow and Schurz into positive cooperation," the newspaper project being conditioned on Bristow's candidacy, he set about trying to engage an editor. Successively he approached Horace White, long connected with the Chicago *Tribune* as an independent journalist, Charles Nordhoff, the recent author of *American Communistic Societies*, Francis A. Walker, who had collaborated with him on "The Legal Tender Act," and finally Carl Schurz, leader of the Liberal Republican movement and senator from Missouri. Schurz was to be paid "a fixed salary of say $5000 a year, and a percentage of the profits over and above a certain sum." [10]

By the end of May, 1876, it became apparent that all this

labor had been wasted. The newspaper scheme evaporated. Bristow fell victim to the caucus. Speaker James G. Blaine emerged momentarily only to be eclipsed by a cloud of scandal connecting him with certain western railways. Only one more straw was needed, the nomination of Rutherford B. Hayes, "a third rate nonentity." Schurz, mindful of the main chance or at any rate more realistic than Adams's coterie, went over to Hayes and left his followers in the lurch.

Once more political idealism had gone down to defeat. In that defeat one piece of strategy must have afforded Adams a bit of satirical amusement. At the height of the political negotiations that spring, his group decided to procure a Harvard LL.D. for Schurz, a move which would presumably guarantee the respectability of their foreign-born champion. Doubtless it would also bind him more closely to his New England sponsors. Adams urged the proposal upon President Eliot and directed Lodge to apply a little pressure from his side. The maneuver worked and the hero of the Independents joined Adams's doctoral candidates at the commencement exercises late in June. It pleased him to have carried his point about Schurz, but his high estimate of the man waned. Masking his own bitterness he consoled Lodge: "You are indeed the one who has the best right to complain for you had the most trouble in forming that rope of sand, the Independent party. I cannot help laughing to think how, after all our labor and after we had by main force created a party for Schurz to lead, he himself, without a word or a single effort to keep his party together, kicked us over in his haste to jump back to the Republicans. . . . Well, we knew what he was! I am not angry with him, but of course his leadership is at an end."

There was ample cause for disappointment. He, as well as Lodge, had driven himself without respite. In their heavy correspondence of the time a half dozen different concerns clamored for attention. The *North American* had to be fed with articles and reviews; the learned doctoral essays on Anglo-Saxon law had to be readied for the press and his own essay completed; his new course entailed the most arduous research and lectures had steadily to be ground out. Success in their political campaign might have emancipated him from the "ways of Boston" and have somehow utilized his demonstrated talents for politics. We can only guess what dreams of glory and power passed through his mind. Political disappointment had come so often and so unfailingly to the members of his family since the Civil War that he did not dare confess to either hope or personal ambition.

He now saw that the party machines and the caucus could not be beaten by the efforts of a few well-intentioned and self-appointed leaders. So long as the public remained indifferent little could be done. Sadly he foresaw that the corrupt party machine would very likely outlive him. "I prefer to leave this greatest of American problems to shrewder heads than mine," ran his disillusioned comment to Lodge. "When the day comes on which it will be considered as disgraceful to be seen in a caucus as to be seen in a gambling house or brothel, then my interest will wake up again and legitimate politics will get a new birth."

The contest between Hayes and Tilden seemed to offer little more than a choice between tweedledee and tweedledum; but he decided to support Tilden. As he ruefully acknowledged to Wells, "After chattering about voting for the best man without regard to party, I cannot well do otherwise. I do this in the expectation that if chosen he will be adroit enough to carry hard money by forcing Republicans to support him, and tariff reform by pure Democratic discipline." On the other hand, he admitted that if Hayes were elected he would support him "with great pleasure." The main point in his view was that the

Independents would have to support the administration in any case, as their influence would be appreciable only if they remained united.

If direct political action was now out of the question, there yet remained the rostrum of the *North American* from which to hurl Parthian epithets against all of their enemies. His spirits rose as he prepared to hold inquest over the national body politic in the October issue. The publishers got wind of one of the articles, "The Independents in the Canvass," the joint production of Henry Adams and his brother Charles Francis, an article which advised Republicans to kick over the party traces. Such an affront to Massachusetts Republicans was too much for the publishers, Messrs. Osgood and Houghton, especially as the finances of the magazine were in a precarious state. They formally disavowed all responsibility for the issue.

In spite of the favorable puffs from Adams's friends and his own efforts to solicit business the magazine had not attracted an adequate number of paying advertisers. By 1875 Henry O. Houghton of the Boston printing firm of Hurd and Houghton had acquired some kind of controlling interest in the house of James Osgood and Company, the publishers of the magazine, because of the threatened insolvency of the publishing house.[11] As the affairs of the *Review* were implicated, Houghton, wishing to economize, proposed to reduce the page rate to authors. Adams indignantly declined the suggestion, preferring to let the magazine "die at once." Osgood then offered to sell out to Adams; but the harried editor had temporized. Now, unexpectedly, the opportunity had arrived to free himself at a stroke of the whole business. Availing himself of the quarrel over the October issue, which he termed "a trifling disagreement," he resigned, leaving the volcanic October issue as his "monument." Helpless to do more, the publishers inserted the following disclaimer: "The editors of the 'North American Review' having retired from its management on account of a difference of opinion with the proprietors as to the political character of this number, the proprietors, rather than cause an indefinite delay in publication, have allowed the number to retain the form which had been given it, without, however, committing the Review to the opinions expressed therein."

The offending "political manifesto," "The 'Independents' in the Canvass," restated point by point the intense convictions of the two crusading brothers as expressed in eight years of fearless public discussion. In every plain-spoken line it showed that they had not deviated from their youthful ambition to "blow up sophistry and jam hard down on morality." [12]

In the impending election, the Independent voter, would have to decide first, whether to support Hayes or Tilden, second, what he must do "whichever party comes to power." The authors carefully discriminated between the "live" issues and the spurious ones. "Reconstruction" was obviously "a stale excitement"; for the Independent movement as a whole was largely a protest against the use of that issue as political camouflage. On such issues as the resumption of specie payment, reform of the civil service, and reduction of the tariffs neither party gave ground for much hope. However, Tilden might prove more useful to the reform element than Hayes. At all events the Republican party needed to be driven into the opposition in order to restore to the Executive the independence it held before Jackson. The Independent voter must therefore constitute a third party, prepared to act with either of the other parties when it might serve his interest but resolute in adherence to the one great end: 'to overcome the tendency of our political system to corruption." To root out the source of the evil, "the relation between the party system and the constitutional system must be reversed."

In the same farewell issue H. V. Boynton's article on "The Whiskey Ring" exposed the political ramifications of the revenue frauds, frauds which had touched the White House retinue

itself in the person of Orville E. Babcock. Secretary of the
Treasury Bristow's efforts to prosecute the criminals had finally
encountered the personal opposition of President Grant. Adams
was sorely tempted himself to "add a page on the political
moral," in order to make the article "tally with the tone of the
number" but decided it would be more diplomatic to ask the
author to do it. A disturbing companion piece to this analysis
was the unsigned article by Charles F. Wingate and Charles
Francis Adams, Jr.,[13] called "The Shattering of the Ring," which
gave in full perspective the shocking story of the Tweed Ring
in New York.

The lead article written by William Henry Trescot, the South
Carolina statesman, "The Southern Question," flung an equally
unbearable challenge into the face of the regular Republicans.
Trescot pleaded that the South should be permitted henceforward to work out its own destiny and reëstablish white
supremacy. The plea was at least as unpalatable as the one
made by John Quincy Adams II in 1868 as Democratic candidate for governor of Massachusetts. It said in effect what
Brooks had conceded in 1874 and what Charles, Henry, and
their father had been saying from the closing days of the Civil
War. Social reform in the South was not the office of the federal
government, nor had it been the chief aim of the North. In any
case events had shown that the federal government had neither the will nor the power to carry out the Draconian system
of Reconstruction projected by Northern extremists.

The October *North American* was indeed "a very strong
number" as assistant editor Lodge reported to the historian
Bancroft in Washington. The printing was immediately exhausted, an unprecedented occurrence in the history of the
magazine. Here once more was advertisement beyond Adams's
hopes, but once again it was unusable. The *Nation* which had
so long approved the magazine, whose reformist policies paralleled its own, reported that "a severe internal convulsion" had

taken place. The reviewer heatedly denounced the censorship which the publishers wished henceforward to impose. The *North American* had been the only "high class" magazine in America in which "a careful political article by a competent thinker and observer could find a place." By forcing the retirement of Adams the publishers had struck a severe blow to the liberal cause.[14]

Emancipation

Just as the October *North American* stood as the "historical monument" to Adams's editing, the *Essays in Anglo-Saxon Law* served as a monument to his teaching. Editing and teaching had made their mark, but politics had died in a nameless grave. The equivocal election of 1876 dropped Adams into an emotional vacuum in which his twin monuments signalized for him the end of an era. The failure of politics left him suspended between the Stone Age of President Grant and the Brazen Age of President Hayes. What Adams's feelings were during his long winter of indecision we can only conjecture, for the published letters suddenly run out at the height of interest, not to resume again until April of 1877.

Assuredly the time had come to take stock of his career. Once, like Disraeli's Coningsby, he had thought, "I should like to be a great man," but like him he had learned that "the history of heroes is the history of youth." He was all too conscious that youth was gone. At thirty-three he had sadly jested, "I am growing old." Already at thirty-nine his head gleamed like that of Chaucer's monk and he was beginning to see himself as "Casaubon in *Middlemarch*." "Aridity grows on me," he acknowledged to Lodge. Beneath the spirited badinage of his brilliant letters to Gaskell, who was now a Member of Parliament, runs a persistent note of regret, the regret of an ardent young man who had taken to heart the portrait of Disraeli's and Bulwer-Lytton's heroes. In London, in the sophisticated company of his friend Gaskell, it had been flatteringly easy to identify himself with the irresistible Coningsby and the exquisite Pelham. Adams like Pelham had pored over Mill's *Essay on Government,* and his family had taught him what Pelham also discovered: "how inseparably allied is the great science of public policy with that of private morality." Had he not also tried to make his life conform to Bulwer-Lytton's purpose and show "how a man of sense can subject the usages of the world to himself instead of being conquered by them"? Had not he and his friend Gaskell once wished to be "both men of the world, and even, to a certain degree, men of pleasure, and yet be something wiser, nobler, better"? Nineteen bewildering years had gone by since he set sail from Queenstown aboard the "China" to seek his fortune in Washington. There was no prospect now of being either a Coningsby or a Pelham, and the names of these youthful literary enthusiasms could survive only in mocking allusions.

Already he had shifted to his scholars' shoulders the courses in which he was no longer interested; in a pinch they could easily take over the rest of his work. Henceforth even graduate teaching would only be repetition, for he had baked his first "batch of Doctors of Philosophy." The vivid pleasure that he had taken in the achievement of his three distinguished students would not come again; they were now the first of an endless file of scholars. He had spent his energies during the past year with prodigality and he was tired. In the fall of 1872 he had returned to his chair in medieval history with relish for the work lying ahead. In the fall of 1876 he unwillingly braced himself for the new year. Eager to base his new course—American History: 1789-1840—upon original documents, he buried himself beneath "avalanches of State Papers" and when he emerged he faced an unexpectedly large class. "I am just beginning my grind at the university wheel," he wrote to Gaskell, "and for my sins I am becoming popular in my old age. My classes are very large, one of them is near seventy in number.[1] As I detest large classes, I am much disgusted at this, and have become foul and abusive in my language, hoping to drive them away." Forced to conduct his new course almost wholly by lecture, a method which he thought defrauded both parties to it, he was bound to find the new schedule irksome.

His discontent was of no sudden origin however, and it ran deeper than a distaste for large classes. A year and a half earlier he had meditated, "marriage makes a man quiet." But quiet, he discovered, did not bring content. The Boston which his brother once described in the *North American* as having lost both its influence and its past was becoming a provincial city in which intellectual life had foundered in moral earnestness. Adams's reaction to this degradation was almost violently chemical. Only gross exaggeration satisfied his feelings. "For twenty-five years, more or less, I have been trying to persuade people that I don't come from Boston," he later wrote to Godkin. He had a horror of growing provincial, of becoming the "most odious" of Boston prigs, an "intellectual prig." He confided his sense of frustration as a teacher to his old friend Francis A. Walker: "Between ourselves the instruction of boys is mean work. It is distinctly weakening to both parties. I have reduced my pedagogic work to the narrowest dimensions and am working back into active life." Yet when he thus wrote to Walker more than a year of "pedagogic work" still lay ahead; and by chance the drudgery was to be redoubled. His was precisely the dilemma that Eliot had tried to avoid when he urged the creation of a graduate school, his argument being that "as long as our faculty regard their work as teaching boys, we can never get first-class instruction." Contrary to Eliot's expectation Adams's graduate class did not sufficiently palliate the drudgery.[2]

The impending election irresistibly drew his thoughts to

Washington. There lay the best hope of escape. "If the new Secretary of State is a friend of mine," he reflected, "I shall try the experiment of passing a winter in Washington." In any case he resolved to keep his hands free and make no new commitments. President Daniel Coit Gilman of Johns Hopkins University enterprisingly used the quarrel over *North American* policy as the occasion to offer Adams a professorship in history in the new university, coupled with an offer to buy the *North American* in his behalf. It was tempting bait, yet to a man nearing thirty-nine, acceptance might well tie him irrevocably to the academic life. Then all the reservations that had surrounded his original acceptance of the Harvard offer would frankly have to be abandoned. Clearly intimidated by the prospect, he replied that if the Baltimoreans should buy the magazine he would be glad to aid the new editor. "But," he added, "I feel so uncertain whether I shall stay quiet more than a short time longer, that I cannot put on harness again." [3]

The uncertainty of the winter continued into the early spring of 1877. His contract at Harvard, however tenuous its legal hold upon him, still had three years to run, and he continued to go through the motions of hard work. Recalling his arguments with Lodge over New England Federalism, one of the subjects of the winter's lectures, he thought of a plan for stimulating the interest of their students in American history. Why not introduce the principle of competition between teachers of the same subject? In a letter to President Eliot he offered to set up a rival course to his own in American history 1789-1840 "to be given by Mr. Lodge." The Corporation "gladly" assented to his bizarre plan. Still his resolve to quit had not reached the sticking place, though his conviction grew that his university work was "essentially done" and nothing remained but "mere railing at the idiocies of a university education." He had been a professor, he intimated to one of his brilliant seniors, Henry Osborn Taylor, "as long as one ought to be." [4]

Within a month his half-formed intentions were finally crystallized by a chance occurrence. Albert Rolaz Gallatin, the son and literary executor of Albert Gallatin, Jefferson's great secretary of the treasury, employed Adams to edit his father's papers. The two men had known each other for years, their acquaintance dating back perhaps as far as 1868.[5] One may guess that George Bancroft, who had already settled in Washington to prosecute his historical researches, had been the intermediary. In any case Gallatin could be satisfied that Adams's current researches in the period 1789-1840 peculiarly qualified him to undertake the work. In April, therefore, he went down to New York, his students presumably left in Lodge's charge, to arrange the "great mass of papers," which he was obviously relieved to discover would "give me some years' work and exercise a good deal of influence on my future movements." With such a tangible and important piece of work in hand, he could now cut loose from Harvard.

It was fortunate for his new career that the political situation in Washington promised not to distract him. William Evarts with whom he had long been on the friendliest terms was now Secretary of State; but as all of the Adamses had publicly supported Tilden, Evarts could offer no more than a private desk and free access to the archives of the State Department. "This," remarked Adams, "is an illustration of the way politics works; always unsatisfactorily." The conclusion was hardly sound, for his family's case was distinctly atypical; they had uncompromisingly fought the party system and had lost. As political heretics in state and national politics, they had to pay the going price of heresy. For himself, Adams was prepared to concede that political management was not his forte. "Long experience has taught me," he once wrote Wells, "that things never turn out in politics as I think they will, and that good generally comes from confusion." Such vague optimism seems strangely out of character. All along he had known that his own chances for high political preferment were slim. In a moment of self-abasement he once acknowledged to Gaskell, "In no case can I come in for any part of the plunder in case of success. My father and brothers block my path fatally, for all three stand far before me in order of promotion." [6]

In a world of political determinism there was no room for the consciously philosophic statesman. He was thus left to hobnob with the leaders of both parties, "very contented under my cloak of historian." He was satisfied, so he said, that "literature offers higher prizes than politics." George Bancroft had made a fortune out of history and had come back to Washington from his long mission to Germany an elder statesman, full of years and honors. Motley, whom he also knew, had likewise prospered and had acquired a vast European reputation in the process. If direct political power was inaccessible to him, he could aspire to reputation and influence comparable to theirs.

Against such an exalted perspective a career as a mere free lance in the press, attacking "corruption in all its holes" had little further attraction. Moreover the matter of reform was no longer desperate. His friend Carl Schurz was in the Cabinet as Secretary of the Interior and could be depended on to foster the limited objectives of their group. Reform no longer called for men to stand alone. It too had entered the orbit of political determinism, having become an organized movement, and was now the occupation of patient journeyman workers. As one of the solitary and eloquent prophets of the movement he could now go his way in peace.

Whatever the "higher" prize he now sought, its shadowy outlines prefigured the intellectual and the artist. His apprenticeship to the muse of history had been protracted and arduous. As if to symbolize his emergence and to make his departure from Harvard irrevocable he gave "all his German books on law and the sources" to his friend Oliver Wendell Holmes, who remained on as teacher of constitutional law, "a noble present" as Holmes later characterized it to Sir Frederick Pollock.[7] Embarking at last on a career as writer and historian, fulfilling after a fashion his undergraduate ambition for a quiet literary life, he now surrendered all thought of a joint career with his brother Charles, whose single-minded devotion to railroad affairs was soon to lead him from the Massachusetts Railroad Commission to the presidency of the Union Pacific Railroad.

It was clear to Henry Adams that the family go-cart, like the fabulous One-Hoss-Shay, had disintegrated; old-fashioned statesmanship had vanished from the American scene. The great age of the machine politician had arrived. Unsuccessfully the Adamses had tried to maintain the amateur status appropriate to hereditary statesmen. They had tried to elevate the electorate with dignified appeals to reason and morality rather than to self or class interest. Confronted with public apathy and even hostility, they seemed not to understand that these were the fruits of the very system which they had immemorially worshipped as the perfection of political design: the system of checks and balances which so skillfully pitted interest against interest that to the mass of men political action seemed unreal and futile. Henry Adams even persuaded himself to believe that the Civil War proved the adaptability of that system; yet he deplored the rise of the ominous twin system—party government—a system dictated by the inescapable realities of economics. Operating outside the constitutional framework, well-financed pressure groups could employ the party machinery to achieve their ends, unhampered by academic checks and balances.

Frightened by what his brother called the new Caesarism of business, the concentration of enormous wealth in a few hands, Henry clung to the family belief that the remedy for the abuse of economic power lay in private morality. Essentially it was a religious conception, for it exhorted each man to reform from within, disregarding the actual system of rewards and punishments collectively decreed by the market place. No such division between individual morality and the morality of commercial society obstructed the progress of the corporate Caesars. The great corporations, though immortal, lived in but one world, and being persons without souls were undistracted by spiritual considerations. Only the Constitution which gave them personality could give them a spiritual existence and a system of morality. To expect the private morality of stockholders, directors, and officers to impose itself upon a creature perfectly designed to seek pecuniary gain, except as limited by the laws which its campaign contributions could alter, was to expect a legislative miracle. Unwilling to apply scientific method to devise an efficient political system for the new industrial order, he was to drift steadily into the camp of the philosophical idealists who denounced the reality which their ideas disabled them from changing.

The unvarying course of this development was evident from the beginning, from the romantic idealism of his Class Day oration at Harvard, through the questionings of his article on Lyell's geological theories, and through the basically conservative preconceptions underlying his political writing. Having from the first been confirmed in his fear of philosophical materialism, a fear which he shared with his friend and one-time teacher, James Russell Lowell, he was cast into a life-long dilemma, well symbolized by his 1894 comment on Karl Marx: "I think I never struck a book which taught me so much, and with which I disagreed so radically in conclusion." His efforts to extricate himself from that dilemma with his idealistic preconceptions intact forms the fascinating and often baffling story of his later writings which were to provide American literature with its most brilliant casuistical treatment of modern science.

A brief obituary to Adams's political career appeared in Warrington's *Pen Portraits* of 1877. The author, William S.

Robinson of the Springfield *Republican,* the one-time partisan of the unforgiving Charles Sumner, wrote: "Mr. Henry Adams had too much of the English, and diplomatic and supercilious character which belongs to the New York *Nation* to allow him to become a useful public man." The gibe, malicious as it was, had its measure of truth. Adams had in fact stood by his vow never to make a political speech. In a world of political evil he would now adopt the maxim of the dervish in *Candide:* "Be silent," for silence would hereafter be best. In a blank book of "Characters" one of his aunts by marriage made the following shrewd appraisal of her nephew, after studying his easily identifiable hand writing: "The man here portrayed holds himself aloof. His normal state is that of the elemental warfare of man with work. He wishes to be a force, and yet takes no part in life's high destiny. He sits like a mountain, aside . . . Advanced in thought, knowing and strong, he sees clearly and he acts mildly . . . He cannot act—he sets himself before himself as the universe." [8]

The decision to leave Harvard and Boston once made, Adams put the past resolutely behind him. Any allusion to it seemed an intrusion, as when President Gilman some six months after his resignation asked him to lecture on his experiences as a teacher. He replied, "I confess I would rather not talk about my own experiences as a teacher. They satisfied me so completely that teaching is and always must be experimental if not empirical, in order to be successful, that I was glad to find an excuse for abandoning any wild ideas I might have had of creating a satisfactory method of pedagogy. My only advice to my scholars who succeeded me in my branches of instruction was: 'Whatever else you do, never neglect trying a new experiment every year.' It was a confession of failure, and all the more because it was intended to stimulate the instructor rather than the student." [9] Gilman must have been impressed by the flash of insight; he must also have puzzled over the Delphic obscurity. Was Adams really being wholly honest with himself? Or was the counsel of perfection one more rationalization of his desire to escape to Washington?

In a letter written to his friend Charles Milnes Gaskell he described the change with a freshness and buoyancy long absent from his correspondence. "We have made a great leap in the world, cut loose at once from all that has occupied us since our return from Europe, and caught new ties and occupations here. The fact is I gravitate to a capital by a primary law of nature. This is the only place in America where society amuses me, or where life offers variety. Here too, I can fancy that we are of use in the world, for we distinctly occupy niches which ought to be filled. As I belong to the class of people who have great faith in this country and who believe that in another century it will be saying in its turn the last word of civilisation, I enjoy the expectation of the coming day, and try to imagine that I am myself, with my fellow *gelehrte* here, the first rays of that great light which is to dazzle and set the world on fire hereafter. Our duties are, perhaps only those of twinkling, and many people here, like little Alice, wonder what we're at. But twinkle for twinkle, I prefer our kind to that of the small politician . . ."

tions of the *History* into a Carlylean prolegomenon to a science of history. But those speculative studies were to have a curiously posthumous character, the result of a fantasy with which he consoled himself for the loss of his wife. After her death in 1885 he rebuffed intruders from beyond his circle with the sardonic apology that he had been dead for many years. The *History,* he insisted, belonged to the "me of 1870, a strangely different being from the *me* of 1890." The light had gone out, he said in his bitterness. "As long as I could make life work I stood by it." Yet neither sorrow nor self-disgust had force to swerve his pen. The fabric of the work showed an unmarred pattern and the final chapters rose to the height of the long-sustained argument with undiminished brilliance.

By 1879 Henry Adams already felt himself a dedicated instrument. The city of Washington was his Rome, the world focus of the democratic and secular spirit as Rome had been of the religious and hieratic spirit. The Capitol was the temple to the dreams of his ancestors, however, and not the tomb. Here was to be written not the decline of a world but the rise and progress of one, not history but prophecy. His work on the Gallatin papers inevitably sent him to the immense Jefferson and Madison collections in the State Department, where his friend Secretary Evarts graciously assigned him a desk and gave him carte blanche to the manuscripts, not without some mutterings of favoritism from other scholars. These collections led him to follow the diplomatic trail abroad for the secret counterparts in the archives of London, Paris, and Madrid. "I want to tell the whole truth in regard to England, France, and Spain, in a 'History of the United States from 1801 to 1815,' which I have been years collecting materials for," he explained to Minister James Russell Lowell, whose help he wanted with the Spanish archives. Far better circumstanced than his model, Gibbon, Adams's researches in the company of his wife were a charmed social adventure. His English and Continental affiliations opened files hitherto bound with triple brass. Few diplomatic secrets resisted his urbane assault. Possessed now of large private means he set amanuenses to work where he could not himself transcribe documents in his own impeccable script.

The early years in Washington were years of ripening literary powers, a period of rich intellectual and social fruitage. Sometimes the turbid cross currents of his life bewildered him. He felt the flight of years and continents; but, in the main, life seemed "real and enjoyable as ever," as he assured Gaskell. "Indeed, if I felt a perfect confidence that my history would be what I would like to make it, this part of life—from forty to fifty—would be all I want. There is a summer-like repose about it." In that genial climate he toiled beneath its sun like a happy bondslave while the ruled legalscap pages of his manuscript rose relentlessly.

The long foreground of his career brought him to the task as by a kind of apprenticeship. Motives more than personal fame urged him on. In his family, American history could not help but be, in part at least, genealogy and even filial piety. The ideal which he had set before the young Henry Cabot Lodge suited his own case more perfectly. Each of the great historians of his acquaintance had made one age or national epoch his own: Bancroft, Motley, Parkman, and Palfrey within the circle of his American friendships; and Maine, Stubbs, Green, and Bryce among his English friends. Added to these impulses was the conviction that the whole American scheme of things was first seriously tested before the world in the period which he had settled upon. His work became therefore a great vindication of America against England—and Europe.

In the biography of Albert Gallatin, the writing of which brought him to Washington in 1877, Adams discerned the difficulties which prevented the application of a priori principles to the art of government. The program of Gallatin and the presidents under whom he served, Jefferson and Madison, was "broad as society itself, and aimed at providing for and guiding the moral and material development of a new era,—a fresh race of men." This had been the hope, but the Jeffersonian system made "too little allowance for human passions and vices." These, erupting into "foreign violence and domestic faction," had prostrated the triumvirate of idealists. Since their day, said Adams, "no statesman has ever appeared with the strength to bend their bow." The *Gallatin* could make only a preliminary statement of the national problem. The implications needed to be worked out in a more philosophical framework. With some success Adams read into the record a progressive national development; however, an aura of disenchantment strangely mists over the seeming upward movement.

In the year in which the *Gallatin* was published, 1879, Adams wrote his first novel, *Democracy*, passing the rapidly written sheets under the scrutiny of his wife, who knew the prototypes of the characters and doubtless helped as a conspirator in the fun of writing. Obviously undertaken as a *jeu d'esprit*, its mordant satire relieved the pent-up opinions which the study of Gallatin's career had brought into historical perspective. Scholarship confirmed what Grant's Washington had shown to his direct observation. Jeffersonian democracy, as vulgarized by Jackson, carried corruption in the bone. Ten years before, Adams had aimed his "blow at democracy," at the evils which seemed to flow from universal manhood suffrage and party government, evils which English critics of democracy like Carlyle, Ruskin, Arnold, Stephen, Maine, and Lecky had so long dinned in his sensitive ears. The blow, doubled and redoubled through the pages of his Washington articles, had glanced harmlessly from the hide of the behemoth. All that residue of feeling, reawakened by the spectacle of Gallatin's futile struggles, demanded outlet, purgation. That in large part must have been the office of the novel which he dashed off in a few months.

Serious as were its limitations of plot and characterizations, the novel taught him much about the art of investing a narrative with drama and movement. The advance in method is clearly reflected in his *John Randolph*, likewise written as a relief from the exacting scholarship of the *History* and out of the debris of materials being left over from the main work as it progressed. In it Adams seemed to pick up where he had left off in his early attack on Captain John Smith and the sacred kine of the Virginia tradition. Unsympathetic and partisan though it is in its interpretation of the character of Randolph, the book is brilliantly sustained in its portrait of an erratic and evil genius. Beside it, the *Gallatin* is pedantic. In artistic form it exhibited all the Adams "instinct for the jugular." Its author was achieving mastery in the art of depreciation, a mastery nourished by the example of such practitioners as Horace Walpole and Junius.

To this period also belongs the strangely missing Aaron Burr manuscript. As Adams drove steadily on through the volumes of the *History*, he saw at every turn tantalizing glimpses of national saints and sinners. Aaron Burr had caught his fancy as one of the more impressive archangels who would sooner reign in hell. Hitherto, Burr had been little more than a mechanical contrivance of treason in American history. He was a historical riddle that challenged solution. What direction Adams's iconoclasm took in the full length study can be guessed from the condensed sketch in the major work. Burr was a symptom, as was the Essex Junto and the Hartford Convention, of fateful tensions in the American system.

One final work along the way refreshed Adams when he reached the end of the first section of his history. Again he turned to the novel and again he published without avowing authorship, this time effecting an almost total withdrawal from the mass of the reading public whose taste he despised. He forbade any kind of advertisement, asserting a wish to make an experiment of public taste. In a sense, this novel, *Esther*, is a transmutation of Adams's studies of the primitive rights of women. In a kind of allegory it expressed thoughts which had long obsessed him. What was the *new* woman to be in America? What would woman's rights do to her personality? Could she, as a mere counterpart of man, be saved to the family which had been the highest creation of her sex? Linked to these questions were the central problems of religion and art. All too soon the novel became a touching memorial to his dead wife who had been a symbol of his remarkable dependence upon women. He vowed that he cared "more for one chapter, or any dozen pages of *Esther* than for the whole history."

Following the examples of Palfrey and Bancroft, he privately printed the provisional first section of the *History* in an edition of six copies with wide margins provided for corrections. These copies went to intimates like John Hay, Carl Schurz, Bancroft, and other trusted critics. By the end of 1888 the second section was ready for preliminary printing. Six copies were again distributed. Then began the exhausting labor of revision and enlargement. The first administration of Jefferson was issued in October of 1889; the final volumes—VII, VIII, and IX—comprising the second administration of Madison were published in January, 1891.

To Adams the critical reception of the work left something to be desired. The book was widely noticed, of course; and for the most part comments were highly favorable. The New York *Critic* believed that the study approached "nearer the standard of science, than any extended historical work on this side of the Atlantic." The book trade was advised by the *Book Buyer* that it left "little or nothing for any subsequent historian to attempt" in the domain which it had preëmpted. Its opening and closing chapters were a "masterly survey." In the *Nation* the former contributor was honored with long two-issue reviews for each of the administrations, and the baffling historical problems which Adams had so skillfully handled were fully rehearsed. Adams's friend Worthington C. Ford told the readers of the *Political Science Quarterly* that the *History* furnished one of those rare opportunities "for bestowing unlimited praise." A violently dissenting voice was that of a mysterious "Housatonic" whose letters to the New York *Tribune* were reprinted as a thirty-four page pamphlet, "A Case of Hereditary Bias." Reviewers generally tempered their praise by challenging many of his interpretations of character and episode and variously reacted to his Olympian intolerance of human weakness. However, the review which he had hoped for —the definitive full dress essay, like the one which he had long ago written on Lyell and like those which made and unmade reputations in the great English quarterlies—such a review was not forthcoming. Even his friend Gabriel Monod of the *Revue Historique,* to which he had supplied an article on Napoleon in 1884, seems to have slighted his volumes. They left no easily followed trace in the important German reviews. A belated notice in the *English Historical Review* dismissed the work as "polemical" and showed resentment at Adams's bias against the statesmen of the Mother Country. These were some of the straws in the wind that refused to blow a gale.

In this respect Adams's self-identification with Gibbon broke down painfully. He had created no sensation. The middle volumes had reached an average sale of 2000 copies in the first year of publication, but the final set of three volumes barely sold 1500 copies in the first year. He could not exult as Gibbon had: "My book was on every table, and almost on every toilette. The historian was crowned by the taste or fashion of the day." The truth was, as the *Atlantic* reviewer put it, that the book had "to encounter the misfortune of having been overmuch expected, since nearly a score of years must have elapsed since it was first whispered abroad that this work was in process of creation." Moreover, his special advantages as "a member of the historic Adams family, presumably steeped in fitness for this especial labor," gave the world "the right to anticipate a great production." That the book "very nearly fulfilled" that anticipation was therefore no matter for astonishment. This kind of complacency must have reminded him, somewhat ruefully, of a prophetic line in one of his letters—that he was writing for readers fifty years hence.

The value of the work as technical history is properly the concern of the specialist. One may note that the standard *A Guide to Historical Literature* terms it "one of the very best pieces of work American historians have produced." It is the style and the philosophic drift that arrest the general reader. An occasional image coruscates with light borrowed from the future: "Chaos had become order," "Society persisted in extending itself in lines which ran into chaos," "The wit of man often lagged behind the active movement of the world." In the pages of this work he achieved what was not again to be encountered in American historiography, a genuinely great style, one that is frankly literary, as was Gibbon's and Macaulay's. Striving for ever greater and greater compression of statement, he evolved a sentence pattern so taut that at times the idea moves with the swiftness of a vibration along a plucked string. Long sentence rhythms everywhere have given way to a strikingly modern tempo: brief but cumulative pulses artfully modulated with antitheses and discreet parallelisms. The long infantry marches of his prose are relieved by apt literary allusions, never esoteric, but commonly to the well-remembered reading of his undergraduate days: Shakespeare, Euripides, Milton, Smollett, and other familiars. But nowhere is the narrative encumbered with ornament.

In conception and execution the *History* was a large-scale application of the technique of interpretation which he had initiated in his Harvard classes and to which he gave preliminary form in the Centennial issue of the *North American Review*. Facts as such had no importance. What was important was the historical pattern which they could be shown to exhibit. Thus he had urged his contributors to the Centennial issue to measure the movement of American society in certain significant categories of the national experience: science, politics, economics, religion, education, and law. This formula, modified somewhat, dictated the framework of the *History*. For this massive study of a lengthened moment in American life was obviously projected as part of a scientific history, a work which might establish a basic datum line in each of the main phases of development. As Adams said in his conclusion: "The scientific interest of American history centered in national character, and in the workings of a society destined to become vast, in which individuals were important chiefly as types." It was as great national types and distinctly not as heroes that Jefferson and Madison peculiarly interested him. The United States, in his view, offered the best laboratory for deducing the laws of a science of history because the economic evolution of a large democracy must be relatively uncomplicated.

In this spirit he assayed the various aspects of the national character in finance, diplomacy, legislation, warfare, tracing as well the accompanying movement of population and wealth in each of the sections. However, it was chiefly the "movement of thought" that interested him, the resolution of a distinctly national character in religion, the arts, and the sciences. He fixed upon the period 1800-1815 as one in which American society was moving toward a new equilibrium, a period of rapidly accelerated changes. In this fifteen-year period, according to his theory, the national character was formed, if not in fact fixed. Parting company with the school represented by his friend Bancroft, he took his stand on the notion borrowed from Comte and Buckle that "the laws of human progress were matter not for dogmatic faith, but for study." As the year 1815 marked the point at which "for the first time Americans ceased to doubt the path they were to follow" and the point at which America's "probable divergence from older societies was also well-defined," it could properly be taken as the fixed point from which the progress of the American system might subsequently be measured.

Flattering as the analysis of the American character was at many points, the whole impact of the book could hardly be described as optimistic. When the current crisis exposed the mysterious fevers of finance capitalism, Adams wrote: "If I thought I should be alive twenty-five years hence, with my full powers of mind and body, I should prepare to continue my history, and show where American democracy is coming out." On a philosophical level he did continue his history in the mathematical allegory of the *Rule of Phase Applied to History*. On the historical level he was to begin again and seek his fixed point in the middle ages and the worship of the Virgin, the attempt of the *Mont-Saint-Michel and Chartres*. In the final chapters of the *Education* he struggled to complete the calculation in terms of the dynamo. What might have been a parallel study of the movement in American history since 1815 was absorbed by the more ambitious project.

Once a believer in human perfectibility, Adams's conception of progress had grown thoroughly equivocal by the end of the ninth volume of the *History*. Progress did not mean the steady improvement suggested by the analogies of genetics and Darwinian evolution. The comparisons to be made with biology were inadequate to represent the transformation which he was studying. Such metaphors had been useful in his researches into early English institutions in a society in which the rate of change was relatively gradual. American society presented him a historical system in which the velocity of change was so rapid that the conception of a developing organism did not fit the facts. Only the new science of Energetics could provide suitable analogies. Herbert Spencer had pointed the way by suggesting that society was essentially mental energy, mind in a state of motion traveling along right lines. Taine had likewise taken up the idea. By turning to the imagery of physics Adams came to envision a society whose success must be judged by the amount of its energies and its ability to control them. Progress in such form could hardly please the taste of cultivated Americans of his class.

As early as 1883 he forecast the theory which became absolute in 1891 upon the publication of the *History.* Writing to Samuel J. Tilden he said: "My own conclusion is that history is simply social development along the lines of weakest resistance, and that in most cases the line of weakest resistance is found as unconsciously by society as by water." In the final chapter of the book, "American Character," he dramatized the hypothesis by describing the course of the Rhine as it flowed from its source in the glacier high in the Alps, down through medieval towns and feudal ruins, until it became, as he said, a highway for modern industry, and at last arrived at a permanent equilibrium in the ocean. "American history seemed to follow the same course. With prehistoric glaciers and medieval feudalism the story had little to do; but from the moment it came within sight of the ocean it acquired interest almost painful. A child could find his way in a river valley, and a hoy could float on the waters of Holland, but science alone could sound the depths of the ocean, measure its currents, foretell its storms, or fix its relations to the system of Nature. In a democratic ocean science could see something ultimate. Man could go no further. The atom might move, but the general equilibrium could not change."

The subsidiary imagery reinforced the main figure. The

theory of history which he was trying to formulate relied not only upon the idea of the conservation of energy—as in his assertion that "the law of physics could easily be applied to politics; force could be converted only into its equivalent force"—but also upon the Second Law of Thermodynamics, certainly by implication. He conceived of society as an energy system in which greater and greater quantities of energy were released, but in which there was a corresponding loss of potential, a process of social entropy. In discussing Madison's difficulties with the Senate, for example, he suggested the evil consequences "if the system continued in the future to lose energy as in the ten years past." Similarly, of a certain Resolution, he wrote that it "marked the highest energy reached by the Eleventh Congress." The War of 1812 also illustrated the cyclical release of energy. "Experience seemed to show that a period of about twelve years measured the beat of the pendulum." Now "the third period of twelve years was ending in a sweep toward still greater energy; and already a child could calculate the result of a few more such returns." The figure of that child, wistful and precocious, calculating the horsepower of the energy system of society would one day be a familiar visitant to the readers of Adams's later writings. Like Wordsworth's infant he here brought the early intimation of the analogy which dominated all of Adams's subsequent thinking.

Appendix

THE WRITINGS OF HENRY ADAMS
1855–1877

The starred items do not appear in James Truslow Adams's "A Bibliography of the Writings of Henry Adams," *Henry Adams* (New York, 1933), pp. 213-229. All of the contributions to *The Harvard Magazine* were unsigned. The unpublished manuscripts listed below are in the Library of Harvard University.

1855

"Holden Chapel," *The Harvard Magazine,* I (May 1855), 210-215.
"Resolutions on the Death of William Gibbons," with George E. Pond, *The Harvard Magazine,* II (December 1855), 46.

1856

"Resolutions on the Death of Hazen Dorr," with Hollis Hunnewell, *The Harvard Magazine,* II (June 1856), 223.
"My Old Room," *The Harvard Magazine,* II (September 1856), 290-297.
"Conquest of Kansas," book notice, *The Harvard Magazine,* II (November 1856), 395-396.
"Paul Fane," book notice, *The Harvard Magazine,* II (December 1856), 440-441.

1857

"Retrospect," *The Harvard Magazine,* III (March 1857), 61-68.
"College Politics," *The Harvard Magazine,* III (May 1857), 141-148.
"Reading in College," *The Harvard Magazine,* III (October 1857), 307-317.
"ΚΑΤΟΙΗΣΙΣ ΚΕΙΛΕΙΑ," *The Harvard Magazine,* III (December 1857), 397-405. Review of Keightley's *Preface to Fairy Mythology.*

1858

"The Cap and Bells," *The Harvard Magazine,* IV (April 1858), 125-132.

* "Saint Paul and Seneca," Bowdoin Prize Dissertations, XIV. Unpublished.

* "Class Life" [printed in the *New England Quarterly,* XIV (December 1941), 683-684].

* "Class Day Oration," Friday, June 25, 1858. Unpublished.

* Contributions to the "Minor Poetry" of Harvard as a Krokodeilos of the Hasty Pudding Club, "Social Alligator and Gridiron 1848-1858." Unpublished.

1859

* "Two letters on a Prussian Gymnasium" [printed in *American Historical Review,* LIII (October 1947), 59-74].

1860

Letters, signed "H. B. A.," in the Boston *Daily Courier:*

    April 30, dated Vienna, April 5

    May 9, dated Venice, April 11 and 13

    June 1, dated Bologna, April 16, 17, and
                         Florence, April 23

    July 6, dated Rome, May 29

    July 10, dated Palermo, June 9

    July 13, dated Naples, June 15

    [J. T. Adams lists one letter for June 29. None appears in the *Advertiser* for that date. *Letters,* I, 59, footnote, omits the letter from Rome. The letters published July 10 and 13 were reprinted with annotations in the *American Historical Review,* vol. 25 (January 1920), pp. 241-255.]

"Letter from Washington," in the Boston *Daily Advertiser,* unsigned:

    * December 7, dated December 4 (See *Letters,* I, 65)

    December 10, dated December 7

    December 13, dated December 10

    December 20, dated December 17

    December 27, dated December 22

1861

"Letter from Washington," in the Boston *Daily Advertiser:*

    January 1, 1861, dated December 28, 1860

    January 11, dated January 7, 1861

    January 15, dated January 11

January 16, dated January 13
January 17, dated January 14
January 22, dated January 18
January 24, dated January 21
January 26, dated January 23
February 2, dated January 31
February 6, dated February 4
February 8, dated February 5
February 11, dated February 7
(As indicated in the text, some contributions were not published; others were expurgated.)

"The Great Secession Winter 1860-1861" [printed in *Proceedings*, Massachusetts Historical Society, 1909-1910, vol. 43, pp. 656-687].

* "The Federal Appointments for Massachusetts," New York *Times*, April 5, 1861. Unsigned.

Letters to the New York *Times* from England, unsigned. (See page 105, note 1.)

    Friday, June 7, "Important from England," dated London, Saturday, May 25

    Friday, June 21, "American Topics in England," dated London, Saturday, June 8

    [Two letters were printed under this heading. Only the second, "From another Correspondent," is attributed to him by Charles I. Glicksberg, "Henry Adams and the Civil War," *Americana*, XXXIII (October 1939), 443 ff.]

    Friday, June 28, "Interesting from England," dated London, Saturday, June 15

    Thursday, July 4, "Affairs at London," dated London, Saturday, June 22

    Monday, July 15, "From London," dated London Saturday, June 29

    Friday, July 19, "Affairs in London," dated London, Saturday, July 6

    ° Friday, July 26 [Fragment], "Why Troops Were Sent to Canada," dated London, Saturday, July 13

    Friday, August 2, "From London," dated London, Saturday, July 20

    Monday, August 12, "Affairs in England," dated London, Saturday, July 27

    Thursday, August 15, "Affairs in England," dated London, Saturday, August 3

    ° Saturday, August 24, "From London," dated London, Saturday, August 10

Friday, September 6, "American Questions in England," dated London, Saturday, August 24

Saturday, September 14, "From London," dated London, Saturday, August 31

Tuesday, September 24, "From London," dated London, Saturday, September 7

Thursday, September 26, "From London," dated London, Saturday, September 14

Tuesday, October 8, "Matters at London," dated London, Saturday, September 21

Sunday, October 13, "A Trip to Leamington," dated Leamington, Friday, September 27

Sunday, October 20, dated Glasgow, Friday, October 4

Monday, October 28, "From London," dated London, Saturday, October 12

Saturday, November 2, "Secession Intrigues in England," dated London, Saturday, October 19

Thursday, November 7, "Affairs in England," dated London, Saturday, October 26, and Sunday, October 20

Monday, November 18, "The Attitude of England," dated London, Saturday, November 2

* Friday, November 22, "The American Question," dated London, Saturday, November 9

* Saturday, November 30, "Important from England," dated London, Saturday, November 16

* Monday, December 9, "The Affair of the 'Nashville,'" dated London, Saturday, November 23

Thursday, December 19, "England and America," dated London, Saturday, November 30

* Wednesday, December 25, "The War Panic in England," dated London, Saturday, December 7

* Monday, December 30, "The Anglo-American Issue," dated London, Saturday, December 14

1862

Letters to the New York *Times,* from England, unsigned.

* Saturday, January 4, "Important from London," dated London, Saturday, December 21, 1861

* Saturday, January 11, "The Suspense in England," dated London, Saturday, December 28, 1861

* Tuesday, January 21, "The Affair of the 'Trent,'" dated London, Saturday, January 4, 1862

"A Visit to Manchester. Extracts from a Private Diary," dated Manchester, November 8, 11, 12, 13, 13 (*sic*), unsigned, Boston

Daily Courier, December 16, 1861. [Reprinted in *American Historical Review,* LI (October 1945), 74-89.]

1863

* Report to William Henry Seward, March 27, 1863. Records of the Department of State, Diplomatic Despatches, Great Britain, vol. 82 [printed in *New England Quarterly,* December 1942].

1867

"Captain John Smith," *North American Review,* CIV (January 1867), 1-30. Unsigned. Reprinted and revised in *Chapters of Erie and Other Essays;* reprinted and further revised in *Historical Essays,* 1891.

"British Finance in 1816," *North American Review,* CIV (April 1867), 354-386. Unsigned. Reprinted and revised in *Chapters of Erie and Other Essays.*

"The Bank of England Restriction," *North American Review,* CV (October 1867), 393-434. Reprinted and revised in *Chapters of Erie and Other Essays;* reprinted and further revised in *Historical Essays.*

1868

Review of Sir Charles Lyell's *Principles of Geology, North American Review,* CVII (October 1868), 465-501.

* "The Argument in the Legal Tender Case," *Nation,* VII (December 17, 1868), 501-502. Signed "H. B. A."

[Note: for ascriptions in the *Nation* see text, p. 188 and note 6.]

1869

"The Session," *North American Review,* CVIII (April 1869), 610-640.

"American Finance, 1865-1869," *Edinburgh Review,* CXXIX (April 1869), 504-533. Unsigned.

"Civil Service Reform," *North American Review,* CIX (October 1869), 443-476. Published as a pamphlet, *Civil Service Reform* by Henry Brooks Adams (Boston: Fields, Osgood and Co., 1869), 35 pp.

* "Men and Things in Washington," *Nation,* IX (November 25, 1869), 454-456. Unsigned. Dated, Washington, November 20, 1869.

* "A Peep into Cabinet Windows," *Nation,* IX (December 12, 1869).

[Note: "A Look Before and After" in the *North American Review,* January 1869, is by J. R. Lowell and not by Adams as surmised in *Letters,* I, 149, note. See *Nation,* January 21, 1869, p. 54.]

1870

* "The Senate and the Executive," *Nation,* X (January 6, 1870).
° "A Political Nuisance," *Nation,* X (January 27, 1870).
* "A Delicate Suggestion," New York *Post,* February 2, 1870, p. 2.
* "Mr. Dawes—President Grant—General Butler," *Nation,* X (February 10, 1870).
"The Legal Tender Act," with Francis A. Walker, *North American Review,* CX (April 1870), 299-327. Reprinted and revised in *Chapters of Erie;* reprinted and further revised in *Historical Essays.*
* "*The North American Review* and the Hon. Elbridge G. Spaulding," *Nation,* X (May 12, 1870). Signed "H. B. A."
"The Session," *North American Review,* CXI (July 1870), 29-62. Reprinted and revised in *Historical Essays.* Reprinted by National Democratic Executive Committee, 1872.
"The New York Gold Conspiracy," *Westminster Review,* XCIV (XXXVIII, n.s.) (October 1870), 411-436. Unsigned. Reprinted and revised in *Chapters of Erie;* reprinted and further revised in *Historical Essays.* [Reprinted in Hicks, *High Finance in the Sixties.*]

1871

Chapters of Erie and Other Essays, by Charles F. Adams and Henry Adams (Boston: James R. Osgood and Co., 1871). Includes the following articles by Henry Adams—all reprints of articles previously published: "The New York Gold Conspiracy," "Captain John Smith," "The Bank of England Restriction," "British Finance in 1816," "The Legal Tender Act." The volume was reprinted in 1886, possibly with additional corrections.

1872

"Harvard College," *North American Review,* CXIV (January 1872), 110-147. Revised and reprinted in *Historical Essays,* 1891.
"Freeman's Historical Essays," *North American Review,* CXIV (January 1872), 193-196. Unsigned.
"Maine's Village Communities," *North American Review,* CXIV (January 1872), 196-199. Unsigned.
"Howells' Their Wedding Journey," *North American Review,* CXIV (April 1872), 444-445. Unsigned.

"King's Mountaineering in the Sierra Nevada," *North American Review*, CXIV (April 1872), 445-448. Unsigned.

"Holland's Recollections of Past Life," *North American Review*, CXIV (April 1872), 448-450. Unsigned.

° Review of Bayard Taylor's *Faust*. Unpublished. (See text, p. 229.)

The Administration—a Radical Indictment! Its Shortcomings. Its Weakness, Stolidity. Thorough Analysis of Grant's and Boutwell's Mental Calibre. No Policy. No Ability (Washington: National Democratic Executive Resident Committee, 1872). Reprint of "The Session" (1869-70). (See above, 1870.)

[The review of Denison's *Letters* attributed to Henry Adams by James Truslow Adams was written by Charles M. Gaskell. See *Letters*, I, 226.]

1874

"Freeman's History of the Norman Conquest," *North American Review*, CXVIII (January 1874), 176-181. Signed "H. A."

"Coulange's Ancient City," *North American Review*, CXVIII (April 1874), 390-397. Unsigned.

"Saturday Review Sketches and Essays," *North American Review*, CXVIII (April 1874), 401-405. Unsigned.

"Sohm's Procédure de la Lex Salica," *North American Review*, CXVIII (April 1874), 416-425. Unsigned.

"Stubb's Constitutional History of England," *North American Review*, CXIX (July 1874), 233-244. Unsigned.

"Kitchin's History of France," *North American Review*, CXIX (October 1874), 442-447. Unsigned.

Syllabus. History II, Political History of Europe from the Tenth to the Fifteenth Century (Cambridge: 1874).

1875

"Parkman's Old Regime in Canada," *North American Review*, CXX (January 1875), 175-179. Unsigned.

"Von Holst's Administration of Andrew Jackson," *North American Review*, CXX (January 1875), 179-185. Unsigned.

"The Quincy Memoirs and Speeches," *North American Review*, CXX (January 1875), 235-236. Unsigned.

"Bancroft's History of the United States," *North American Review*, CXX (April 1875), 424-432. Unsigned.

[Adams was responsible only for the "diplomatic" section. See *Henry Adams and His Friends*, p. 129.]

"Maine's Early History of Institutions," *North American Review*, CXX (April 1875), 432-438. Unsigned.

"Palgrave's Poems," *North American Review,* CXX (April 1875), 438-444. Unsigned.

"Green's Short History of the English People," *North American Review,* CXXI (July 1875), 216-224. Unsigned.

"Tennyson's Queen Mary," *North American Review,* CXXI (October 1875), 422-429. Unsigned.

"Palfrey's History of New England," *North American Review,* CXXI (October 1875), 473-480. Unsigned.

1876

• "Ticknor's Life and Letters," *North American Review,* CXXIII (July 1876), 210-215. Unsigned. (See *Letters,* I, 286, 287.)

"Von Holst's History of the United States," *North American Review,* CXXIII (October 1876), 328-361. Signed, Henry Adams and Henry Cabot Lodge.

"The Independents in the Canvass," with Charles Francis Adams, Jr., *North American Review,* CXXIII (October 1876), 426-467. Unsigned. (See *Letters,* I, 287.)

Essays in Anglo-Saxon Law, ed. by Henry Adams (Boston: Little, Brown and Co., 1876). The first essay, "Anglo-Saxon Courts of Law," is by Henry Adams.

Lowell Lecture: "Primitive Rights of Women." Delivered, December 9, 1876. Published in revised form in *Historical Essays,* 1891.

> [The following additional ascriptions are made by Cushing in his *Index to the North American Review:*
> 1. "Denison's Letters and other Writings," vol. 114, p. 426.
> 2. "Dr. Clarke's Sex in Education," vol. 118.
> 3. "Clarke's Building of a Brain," vol. 120, p. 185.
> 4. "Frothingham's Transcendentalism," vol. 123, p. 468. (See *Letters,* I, 287.)
> 5. "Lathrop's Study of Hawthorne," vol. 123, p. 478.
> 6. Review of Vol. II Stubbs, *Constitutional History,* vol. 123, pp. 161-165.

With the exception of Item 2, all of these are crossed out in Henry Adams's copy of the *Index* in the Massachusetts Historical Society. The review of Denison was by Gaskell. See *Letters,* I, 226. It is quite unlikely that Item 2 is by Henry Adams. As he was not responsible for getting out the January, 1874, issue, he would hardly have been tempted by Clarke's book. He was moved, however, to write a review of Freeman for that issue. See *Letters,* I, 257. Also crossed out is a review of Ticknor's *Life, Letters and Journals;* but that is clearly an error. See *Letters*, I, 286, 287. The authorship of the review of Lathrop's book was a closely guarded secret. Possibly Adams *was* its author. See allusion to the review in *Letters*, I, 289. The review of volume II of Stubbs's work was written by Professor W. F. Allen. See William Francis Allen, *Monographs and Essays* (Boston: 1890), p. 364.]

1877

* Review of Lodge's *Life and Letters of George Cabot, Nation*, vol. 25, July 5, 1877. (*Letters*, I, 301, note. Also *General Index to the Nation* 1865-1880.)

Documents Relating to New England Federalism, 1800-1815, edited by Henry Adams (Boston: Little, Brown and Co., 1877).

* Note on Clarence King, *Nation*, vol. 25 (August 30, 1877), p. 137.

NOTES

Bibliographical Note

The full titles of all works cited in the text or in the notes by short title are given in the Bibliography with the customary facts of publication. As the chief sources of biographical information have been the *Letters of Henry Adams 1858-1891* and *A Cycle of Adams Letters 1861-1865*, they will be referred to hereafter as *Letters*, I, and *A Cycle*. The overriding consideration has been to keep the notes within a reasonable compass and to avoid excessive documentation. Wherever possible, I have collected all the authorities referred to in a paragraph into a single note. I have ordinarily not given the references for the large number of quotations from the *Letters* or from *A Cycle* where the approximate time of each quotation is apparent. Since both of those works are arranged chronologically, the full context of a quotation can readily be found. I wish to caution the reader that I have frequently omitted an initial or final phrase without using ellipses (. . .) to indicate the omission, in order to incorporate the quotations into a running narrative. For the same reason I have often changed initial small letters to capital letters or reduced initial capitals to small letters. In all other respects all quotations are made with literal exactness.

I have assumed that the reader is familiar with *The Education of Henry Adams* and that his copy is readily accessible. The treatment of the historical background for the period may strike some persons as excessively meager. My justification is that the period has been one of the most thoroughly written about in American history, Arthur M. Schlesinger, Jr.'s *The Age of Jackson* being the latest and most satisfying introduction to it. For general guidance in the labyrinth of historical details, I have relied on such works as James Ford Rhodes, *History of the United States from the Compromise of 1850 to 1877* and Edward Channing, *A History of the United States,* as supplemented by more recent special studies like Arthur Charles Cole's *The Irrepressible Conflict 1850-1865,* George Fort Milton's

The Eve of Conflict, and Avery Craven's *The Coming of the Civil War* and by biographies like Allan Nevin's *Hamilton Fish.* These I have gratefully listed in the Bibliography together with the works to which I am more directly indebted.

A recent genealogy of the Adamses, including brief biographical sketches, has been compiled by J. Gardner Bartlett, *Henry Adams of Somersetshire, England and Braintree, Massachusetts.* It should also be added that the subject of the present study was christened Henry Brooks Adams, but dropped the "Brooks" in the fall of 1870, for reasons which can only be conjectured: he may have wished to avoid being confused with his younger brother Brooks, or to symbolize the change in his career or simply to gratify an esthetic impulse. The July 1870 *North American Review* carried his full signature. In October the Harvard *College Records* (XI, 243, 251) also referred to him by his full name; later entries identify him as Henry Adams. On October 31 as editor of the *North American* he solicited an article from Jacob Dolson Cox over the signature "Henry Adams."

CHAPTER ONE: BRAHMIN PATTERN 1838-1858

The Boston Standard of Thought

1. C. F. Adams, Jr., *An Autobiography,* pp. 8ff.; *Boston Almanac* (1849), p. 57.

2. Bacon, *Boston,* p. 31; C. F. Adams, Jr., *Autobiography,* p. 21.

3. *Letters,* I, 136; Holmes, *Tercentenary History, passim,* describes the course of study and gives some information concerning Master Dixwell.

4. It is noteworthy, however, that each of Peter Chardon Brooks's sons-in-law publicly expressed his esteem for him upon his death. Nathaniel Frothingham preached the funeral sermon; Charles Francis Adams wrote the obituary in the *Christian Register;* Edward Everett contributed a biography of him to Hunt's *Lives of American Merchants* (New York, 1858), vol. I. The legal complication reflected a long-standing conflict between the taxpayers and the insatiable city government. As Mayor Josiah Quincy admitted in his 1849 inaugural address the personal property tax was so exorbitant that "great numbers of citizens" took refuge in the country "at the annual period of taxation." Among the countermeasures adopted to stop that exodus was the ruling of the School Committee "that no child be admitted to the public schools of the City of Boston, whose parents, if living, do not reside, or pay taxes in said City." Apparently these were not wholly effective for resentful citizens, like Henry's father, seem to have preferred the penalty.

5. Koren, *Boston 1822-1922*, p. 29; "Rules of the School Committee and the Regulations of the Public Schools of the City of Boston," 1849, City Document No. 49, Chap. 6, Sec. 8; C. F. Adams, Jr., *Autobiography*, p. 22.

6. Henry Adams, "Two Letters on a German Gymnasium," printed in *American Historical Review*, LIII (October 1947), 59-74.

7. J. Q. Adams, *Memoirs* (August 2, 1840), X, 345. Emerson admired the integrity and relished in his own way the personal force of the turbulent old man. He wrote: "He is no literary old gentleman, but a bruiser, and loves the *mêlée . . .* He is an old *roué* who cannot live on slops, but must have sulphuric acid in his tea." *Journals*, VI, 349.

8. Charles Francis Adams became the editor and one of the owners of the organ of the "Conscience" Whigs in 1846. He left the paper in 1848. C. F. Adams, Jr., *Charles Francis Adams*, pp. 50ff., 87, 88.

College Justice

1. Morison, *Three Centuries of Harvard*, p. 294.

2. Anderson, *Letters and Journals of General Nicholas Longworth Anderson*. This is the chief source of information concerning the everyday life of Henry Adams as an undergraduate. The introduction supplies much valuable data about the course of study. Crowninshield, *A Private Journal 1856-1858*, contains a large number of allusions to Adams and provides a valuable supplement to the more ample entries in Anderson's letters and journals.

3. Moses King, and T. P. Ivy, *Harvard and Its Surroundings* (Cambridge, 1878), p. 56; J. Q. Adams, *Memoirs* (July 31, 1826), VII, 138; College Records XI, Harvard Archives.

4. White, J. C., "An Undergraduate's Diary," *Harvard Graduates Magazine*, XXI (March 1913), 423; Kellen, "Winslow Warren: A Memoir," *Proceedings*, Massachusetts Historical Society 1930-1932, vol. 64, p. 53.

5. Morison, *Three Centuries*, p. 287.

6. The Greek tutor, Evangelinus Sophocles, expressed his own annoyance at the innovation by burning "all the blue books unread." Morison, *Three Centuries*, p. 299.

7. Voorhies, ed., Phi Beta Kappa General Catalogue; *Catalogue of the Harvard Chapter of Phi Beta Kappa*. Henry's two elder brothers likewise achieved honorary membership. Charles Francis, Jr., in 1879 (H. U. 1856) and John Quincy (II) in 1884 (H. U. 1853). Only President John Quincy Adams (H. U. 1787) was elected in course. President John Adams (H. U. 1755) appears to have been passed over.

The Mental Scale of Harvard

1. *Education,* p. 34; *Letters,* II, 563, 332, 544, 546, and *Letters to a Niece.*

2. White, "An Undergraduate's Diary," p. 644; Agassiz, *Methods of Study,* p. iv; *Principles of Zoology* (Boston, 1848), Part I, p. 206.

3. Asa Gray in *Dictionary of American Biography.*

4. *Darwiniana: Essays and Reviews Pertaining to Darwinism* (New York, 1876), p. 133.

5. White, "An Undergraduate's Diary."

6. Jouffroy, *Introduction to Ethics* (Boston, 1840), I, 83, 261, 272.

7. Bowen, *Principles of Political Economy* (Boston, 1856), p. 544. For the succeeding quotations from Bowen see pp. ix, 77, 356, 17, 18, 19.

8. The Suffolk Bank of Boston put his professions to the test when in the course of its quarrel with the country banks it sent $40,000 in bills to the Mount Wollaston Bank. According to the New York *Semi-Weekly Times* of October 5, 1858, the specie was "as coolly counted out as if the demand had been for Quincy granite." For an account of the antagonism inspired by the Suffolk Bank System see Justin Winsor's *Memorial History of Boston* (Boston, 1880-81), IV, 161.

9. Foner, *History of the Labor Movement in the United States* (New York, 1947), pp. 367, 413; Johns Hopkins University *Studies in Historical and Political Science,* II (1884), 51 ff.

10. Thomas Arnold, *Introductory Lectures on Modern History,* 4th ed. (London, 1849), p. 28; cf. also p. 151 "The whole character of a nation may be influenced by its geology and physical geography"; *History of the United States,* IX, 224.

11. Robertson, *Reign of Charles V,* p. 315.

12. Guizot, *History of the Origin of Representative Government in Europe* (London, 1852), p. 11.

13. Bowen, *Principles of Metaphysical and Ethical Science* (Boston, 1855), p. 399.

14. Cf. John Adams's remark to his wife concerning his candidacy for the Presidency: "I have a pious and philosophical resignation to the voice of the people in this case, which is the voice of God." January 20, 1796. *Works,* I, 485; Guizot, *Representative Government,* p. 440.

15. Cf. Lowell's later acknowledgment (1878): "I hate it [science] as a savage does writing, because he fears it will hurt him somehow; but I have a great respect for Mr. Darwin as almost the

only perfectly disinterested lover of truth I ever encountered." *Letters,* ed. Norton, II, 230.

Of Men and Books

1. The information concerning the courses of instruction given in the *Catalogue of the Officers and Students of Harvard University* for the years 1854-1858 differs somewhat from that supplied by the instructors' grade sheets. Wherever there is a discrepancy, I have followed the latter. For a tabular analysis of the courses see *Letters and Journals of General Nicholas Longworth Anderson,* p. 20.

2. The catalogue is reproduced by Max I. Baym in *Colophon,* Vol. 3, No. 4 (Autumn, 1938), pp. 483-489, from the manuscript at the Massachusetts Historical Society. Upon his death Adams bequeathed his large library to the Massachusetts Historical Society where it is now stored. This uncatalogued collection of several thousand volumes includes many of the books listed in 1858. The collection had been reduced by several gifts to Western Reserve University, totalling some 250 titles (See Paul H. Bixler, "A Note on Henry Adams," *Colophon,* 1934, Vol. 5, Part 17.), and by the gift of a collection of German law books to Oliver Wendell Holmes, Jr.

3. See *Letters and Journals of General Nicholas Longworth Anderson* and Crowninshield, *passim,* for allusions to current reading of their circle.

Cambridge Life and Letters

1. Class Life of 1858 and *Letters,* I, 5; quoted by George M. Elsey from the Secretary's Records, September 28, 1858, in "The First Education of Henry Adams," *New England Quarterly,* XIV (December 1941), 681.

2. *Letters and Journals of General Nicholas Longworth Anderson,* p. 112. The quotations in this section, not otherwise acknowledged, are almost all from Anderson's *Letters and Journals.* An amusing sidelight on the social activities of Hollis Hall is thrown by the following missive brought to Anderson by Henry Adams:

"My Dear Friend,

A small social gathering is now going on in the rooms of Mr. Charles Adams of the Junior Class. Knowing your talents for conviviality, we should be delighted to have you join us. If you have any brandy or gin, we would be glad if you would bring it along. Please bring some crackers, cheese, or any other delicacies which you may have. We should be pleased to see any of your friends if they are virtuous and agreeable.

God save the Freshman class and the Commonwealth of Massachusetts.

H. Barlow	C. Clapp
C. F. Adams, Jr.	G. O. Holyoke

Noble Mr. Anderson.

(Witness) A true copy.

H. B. Adams."

3. C. F. Adams, Jr., *Charles Francis Adams*, p. 102.

4. J. Q. Adams, *Memoirs*, XII, p. 24; *Charles Francis Adams*, p. 104; cf. *Letters*, I, 57. ". . . not the first time we've been in a minority of one."

5. Morison and Commager, *Growth of the American Republic*, I, 624. The ill-natured debate over the extent and consequences of this assault and battery still goes on between Northern and Southern sympathizers. See, for example, Laura A. White, "Charles Sumner and the Crisis of 1860-61," in *Essays in Honor of William E. Dodd*, especially pp. 133-136.

6. *Letters*, I, 5.

The Harvard Magazine

1. According to Worthington C. Ford, *Letters*, I, vi; "The Tragedy of Mrs. Henry Adams," *New England Quarterly*, XI, 564.

2. *Letters*, I, 19.

3. Adams's meticulously kept set of his own contributions, some transcribed by hand, were bound in a single volume and now rest among his books at the Massachusetts Historical Society. All told, he contributed eleven pieces, amounting to some fifty-five pages. These included two very brief obituaries.

4. The ground had been broken for Adams by two earlier articles in the *Harvard Magazine*, one by James B. Greenough in April, 1856, attacking the Greek letter societies and one by J. C. Ropes in the next issue defending them. A month before Adams's article appeared the senior class voted that the societies should be dissolved.

The Pattern of Success

1. J. Q. Adams, *Dermot MacMorrogh* (Boston, 1832), p. v.

CHAPTER TWO: THE GRAND TOUR 1858-1860

The Pierian Springs of the Civil Law

1. C. F. Adams, Jr., *Charles Francis Adams*, p. 105; New York *Semi-Weekly Times*, September 24, 1858; Boston *Transcript*, November 3, 1858.

2. *Letters,* I, 3; for the precursors of that "deputation," see O. W. Long, *Literary Pioneers.*

3. *Centennial History of the Harvard Law School,* p. 75. A course in Civil Law was offered in 1848-49 and 1850-51.

4. Crowninshield, "Private Journal." See description in Note 5.

5. From an unpublished "Private Journal" of Benjamin W. Crowninshield "commenced at Hanover October 20th, 1858" in the possession of his son Francis B. Crowninshield, who has very kindly allowed me to use the translation prepared at his direction. This record was begun in English but as its author became more proficient in German he changed to that language.

6. January 28, 1859. *Henry Adams and His Friends,* p. 5.

7. Charles Francis Adams, ed., *Letters of Mrs. Adams* (Boston, 1848), p. xxxiii.

8. Adams to Sumner, *Henry Adams and His Friends,* pp. 1, 2, 5, 6.

A Pleasant Series of Letters

1. An opinion which had enjoyed great vogue with foreign travelers ever since Queen Hortense had written in her memoirs that "Tuscany was said to be the happiest state in Italy." Mowat, *States of Europe,* p. 304.

2. King, *History of Italian Unity* (London, 1899), II, 88, and *passim; Letters,* I, 94; Marraro, *American Opinion* (New York, 1932), pp. 282 ff.; Trevelyan, *Garibaldi,* p. 64.

3. Quoted by Taylor, *Italian Problem,* p. 26.

4. New York, *Semi-Weekly Times,* August 27, 1858.

CHAPTER THREE: WASHINGTON CORRESPONDENT
1860-1861

Disciples of the Massachusetts School

1. Charles Francis Adams, Jr., in *Proceedings,* Massachusetts Historical Society, Vol. 43 (1909-10), p. 657; Wallace, *History of South Carolina,* III, 151; Glover, *Immediate Pre-Civil War Compromise Efforts,* p. 33; Boston *Advertiser,* December 7, 1860.

2. J. Q. Adams, *Writings,* I, 70; John Adams, *Works,* I, 595; C. F. Adams, *Address on . . . Seward,* p. 41; *A Cycle,* I, 68.

3. De Tocqueville, pp. 37, 38, 43; *A Cycle,* I, 152.

4. Proceedings, *Massachusetts Historical Society,* Vol. 43, p. 305; John Adams, *Works,* VI, 341; C. F. Adams, Jr., *Charles Francis Adams,* p. 69; Edward Everett to A. H. Everett, October 30, 1846, Massachusetts Historical Society.

4. Adams, *Dana,* II, 256.

5. C. F. Adams, Jr., *An Autobiography*, p. 53; *Congressional Globe*, 1859-60, 36th Congress, 1st Session, 2514, 2516.

6. June 8, 1860. Quoted in White, "Charles Sumner and the Crisis of 1860-61."

7. C. F. Adams, *Address on . . . Seward*, pp. 41, 43; Ware, *Political Opinion in Massachusetts*, p. 26; C. F. Adams (Sr. and Jr.), "Campaigning with Seward in September, 1860," *Minnesota History*, VIII, No. 2, 150 ff.

8. He reached Washington on December 1, 1860. *Proceedings, Massachusetts Historical Society*, Vol. 43, p. 657.

In the Field of Grain

1. C. F. Adams, Jr., *Autobiography*, p. 69. For a scholarly and exhaustively documented account of the confusing struggle between the various factions of the Republican party, see White, "Charles Sumner and the Crisis of 1860-61."

2. C. F. Adams, Jr., *Charles Francis Adams*, p. 140.

3. Morse, *Holmes*, II, 154.

4. *Globe*, 36th Congress, 2d Session, II, 124 ff.

5. C. F. Adams, Jr., *Charles Francis Adams*, p. 144.

6. *Letters*, I, 84. No letter appeared, for example, between January 26 and February 2.

The Great Secession Winter

1. February 23, March 7, March 15.

2. Remained in manuscript until 1910. Printed in *Proceedings, Massachusetts Historical Society*, 1909-10, Vol. 43, pp. 656-687.

3. C. F. Adams, Jr., *An Autobiography*, p. 107.

4. April 1, 1861, *Henry Adams and His Friends*, p. 6.

CHAPTER FOUR: LONDON CORRESPONDENT 1861

Outside of the Legation

1. Frothingham, *Everett*, p. 430.

2. *A Cycle*, I, 12. Cf. also E. D. Adams, *Great Britain and the American Civil War*, I, 98, quoting from Minister Adams's report to Seward, May 12, 1861.

3. *Letters*, I, 85; Gurowski, *Diary* (March 4, 1861–November 12, 1862), p. 59. For similar sentiments see pp. 31, 161, 163.

4. This unpublished diary gives a very interesting if one-sided picture of the daily life of the American legation, especially during the Adams mission. Unfortunately, the minister had no talent for geniality. Even when he intended to be kindly his manner had such "a lack of warmth and a stiffness about it," as Carl Schurz noted

when he called at the Legation, that one felt "as though the temperature of the room had dropped several degrees." *Reminiscences of Carl Schurz,* II, 245. Moran thoroughly disliked his chief and continually belittled his character and work, at least in the privacy of his diary. The manuscript is in the Library of Congress.

5. His intimate friends—Anderson, Cabot, and Crowninshield—became majors, Anderson being brevetted major general for "distinguished gallantry at Chickamauga." Young Lee, another classmate, became a ranking major general in the Confederate army. Davis, *Report of the Class of 1858,* p. 10.

6. Nicolay and Hay, *Lincoln,* IV, 270 ff. The dispatch arrived at the Legation on June 10, 1861. *Letters,* I, 93.

7. Mowat, *Diplomatic Relations,* p. 173.

War of Nerves

1. The full list of them is given in the Appendix. Two dispatches attributed to Henry Adams by James Truslow Adams, *Henry Adams,* those of June 3 and 17, are probably not by him. Plainly written by the same person, they contain a number of allusions inapplicable to Henry Adams's activities, and the adverse criticism of Parliament does not gibe with the Adamses' high opinion of that body. Cf. *A Cycle,* I, 8. Cf. also Charles I. Glicksberg in *Americana,* October, 1939 (Vol. xxxiii), pp. 443 ff.

Extracts from a Private Diary

1. The article has been reprinted in the *American Historical Review,* Vol. 51 (October 1945), pp. 74-89, with a well-informed commentary on the cotton supply question by Arthur W. Silver.

2. Owsley, *King Cotton Diplomacy,* pp. 146-149.

3. January 10, 1862.

4. Thurlow Weed Papers, University of Rochester.

5. *Letters,* I, 206.

CHAPTER FIVE: A GOLDEN TIME 1861-1868

Young England, Young Europe

1. Griscom, *Diplomatically Speaking,* p. 38.

2. Watterson, '*Marse Henry,*' II, 33. "In manners, tone, and cast of thought he was English—delightfully English—though he cultivated the cosmopolite," *ibid.,* 34; "I was impressed with his Anglicized accent . . ." Holt, *Garrulities,* 137.

3. This Date Book is in the Henry Adams library at the Massachusetts Historical Society; Conway, *Autobiography,* I, 406.

4. Walling, ed., *Diaries of John Bright,* p. 263. An allusion to

this conference occurs in a letter of Bright to Sumner. *Proceedings, Massachusetts Historical Society,* Vol. 46, p. 103. An equally scant allusion to Henry Adams has been found in the unpublished diary of William Evarts. The entry, quoted by H. D. Cater in his *Henry Adams and His Friends,* p. cx, reads as follows (London, 1863): "May 19. Breakfasted with Mr. Adams, Mr. Bright, Robert Y. Walker, Mr. Lucas, editor of the *Star,* Mr. Dyer, the traveller in the U.S., Mr. Evans and Mr. Hy Adams, and Mr. and Mrs. Blutchford [Possibly a misreading for *Blatchford.* See *A Cycle* II, 48.] made up the company."

5. *Letters,* I, 99. Motley's evidence seems better than Adams's. Cf. Motley, *Correspondence,* II, 265 ff. The phrase was adhesive. See his re-use of it in the *North American Review* (April 1874), p. 404, and of course the *Education,* p. 200.

6. Thayer, *John Hay,* I, 280 (July 1867).

The New Science

1. Lyell to George Ticknor, November 10, 1861, "I like what I see of Adams, the new Minister here of the United States." *Charles Lyell,* II, 351; in the autobiographical chapter included in *Life and Letters of Charles Darwin,* I, 59; Huxley, *Huxley,* pp. 192 ff.

2. Reprinted in Gray, *Darwiniana,* pp. 109, 169.

3. Walling, ed., *Bright,* p. 264; Jacks, *Brooke,* I, 146. Note allusion to Green and Woolner as fellow geologists, *ibid.,* 233, 237.

4. Jacks, *Brooke,* I, 136. "But my geology is rapidly drifting away from me, and yet what a glorious science it is—while I plunge deeper into historical research." Green in *Letters of John Richard Green,* ed., Stephen, p. 86. Later at Harvard, Henry Adams talked much in his classes of "pteraspis in Siluria." J. L. Laughlin, "Some Recollections of Henry Adams," *Scribner's,* Vol. 69 (May 1921), p. 579.

5. Harrison, *Autobiographic Memoirs,* I, 262; C. F. Adams, Jr., *An Autobiography,* p. 179; *The Tendency of History,* alluding to the church, the state, property, and labor; Comte, *Positive Philosophy,* 8.

6. Spencer, *First Principles,* pp. 271, 380, 396, 428, 484, 506, especially p. 529.

7. In the concluding chapter of the *History.* Cf. Kraus, *A History of American History,* p. 334.

8. Buckle, *History of Civilization in England,* pp. 6, 16, 28. Adams assumed in his later years (*Tendency of History*) that he had read the first volume of Buckle when it appeared in 1857. It seems more likely that he did not read it until after he came to England in 1861. His article "Reading in College," written late in

1857, makes no reference to Buckle; nor is it mentioned in the 1858 catalogue of his personal library. The two-volume edition in his library dates from 1864.

Political Acolyte

1. Harrison, *Memoirs*, I, 265. He had been a lecturer at Maurice's Workingmen's College. Although Henry Adams was won over to the free trade doctrines of the Manchester School of economics, he successfully resisted the economic heresies of Harrison's group. That practical reformer shrewdly observed: " 'Men cannot alter the laws of supply and demand by combination,' they say, in the face of hundreds of successful strikes to raise wages." *Ibid.*, 253. Cf. Henry Adams's dispatch to the New York *Times*, August 12, 1861, concerning the failure of the mason's strike. The men "by creating an extensive system of united trade and mutual aid societies have tried to force up the price of labor. It is a vain hope . . . Meanwhile they earn nothing, and the expense of the effort will more than make up for their gain, even if they were to succeed."

2. Mill, *Considerations on Representative Government* (New York, 1882), pp. 37, 39.

3. *Works*, VI, 399; III, 456.

4. Mill, *Considerations on Representative Government*, pp. 42, 24, 159, 160, 23. Cf. also, "they who can succeed in creating a general persuasion that a certain form of government, or social fact of any kind, deserves to be preferred, have made nearly the most important step which can possibly be taken toward ranging the powers of society on its side." *Ibid.*, p. 23.

5. Editorial comment, John Adams, *Works*, VI, 408.

6. Mill, *Dissertations and Discussions* (1861), I, 260, 407, 197. Note also the long review of de Tocqueville in vol. II.

7. Guizot, *History of Civilization in Europe* (New York, 1887), p. 304; Mill, *Considerations on Representative Government*, p. 138; de Tocqueville, *Democracy*, p. 332; Adams, *The Education*, p. 418, and similarly pp. 100, 147, 365, 503. Lord Acton's subsequent formulation of the axiom "All power corrupts, and absolute power corrupts absolutely" seems now to be taken as the classical statement.

8. de Tocqueville, *Democracy*, pp. 417, 418.

9. Manuscript in the National Archives, Washington, D. C. Printed in the *New England Quarterly*, December 1942.

10. Seward to Adams, April 13, 1863. National Archives, Washington, D. C.

11. *A Cycle*, I, 185.

12. Cf. his brother's comment in 1863: "Your mind has become morbid and is in a bad way . . . My advice to you is to wait until

you can honorably leave your post and then make a bolt into the wilderness, go to sea before the mast, volunteer for a campaign in Italy, or do anything singularly foolish and exposing you to uncalled for hardship . . . I tell you I know you and I have tried the experiment on myself, and I here suggest what you most need, and what you will never be a man without." *A Cycle*, I, 239, 240.

13. Seward to C. F. Adams, August 9, 1864; Henry Adams to Seward, August 25, 1864. On the same date Henry's father wrote Seward that his son would be glad to act in an unofficial capacity until an official successor could be appointed. Records of the Foreign Service, National Archives.

14. The formal letter of resignation is dated December 1, 1867. United States, *Diplomatic Correspondence* (1869), Part I, p. 134.

Path to Power

1. See the series of letters to Palfrey in *Henry Adams and His Friends*, pp. 8 ff. "All readers of the *North American Review* know that the books placed at the head of an article serve often, like the preacher's text, merely to suggest the subject of the article." Cushing, *Index to the North American Review*, p. iii.

2. The fullest atonement seems ultimately to have been made by Charles Francis Adams, Jr. See, for example, "The Ethics of Secession" (1902), in C. F. Adams, Jr., *Studies Military and Diplomatic* and *Lee's Centennial* (1907). On his "recantation" of his early attitude toward slavery, see Bryan, *Joseph Bryan* (Richmond, 1935), p. 370. Cf. also Eliot to C. F. Adams, Jr., April 26, 1914: "You and I are about the same age and began life with much the same set of ideas about freedom and democracy. But you have seen fit to abandon the principles and doctrines of our youth, while I have not . . ." James, *Eliot*, p. 238.

3. Adams to Deane, *Henry Adams and His Friends*, p. 36; Winsor, *Charles Deane*, p. 19, alluding to a comment in the *Magazine of American History*, XIII, 1885. The manuscripts of a few of Adams's articles are among the Norton Papers at Harvard.

4. The rehabilitation of Smith's character continues to be a popular and patriotic historical exercise. See, for example, Morse, "John Smith and His Critics: A Chapter in Colonial Historiography," *Journal of Southern History*, I (May 1935) 123-137. See also *Proceedings*, Virginia Historical Society, 1882, p. 29; Poindexter, *Captain John Smith and His Critics* (Richmond, 1893), pp. 62, 74; Arthur G. Bradley, *Travels and Works of Captain John Smith* (Edinburgh, 1910), p. 99. Marjory S. Douglas, "An Earlier Pocahontas" in *The Everglades* (New York, 1947).

5. Charles Francis Adams, *Reflections upon the Present State of*

the Currency, 13; Smith, *The Wealth of Nations,* p. 883; Mill, *Political Economy,* II, 64.

6. Opinion is sharply divided on his ability as a writer on finance. James Lawrence Laughlin, a noted economist and a former student of Adams, has written that "some of his best work was done in economics in his earlier years." *Scribner's,* Vol. 69, p. 584. James Truslow Adams declares that finance was "always Adams's weakest subject." *Henry Adams,* p. 110. In time Adams became passionately addicted to the study of foreign exchanges and the movement of bullion in his search for a theory of history. See his letters to the late Worthington Chauncey Ford, onetime chief of the United States Bureau of Statistics, in *Letters,* II. The bulk of their correspondence, filled with long tables of statistics, remains unpublished.

7. The Index of the *Pall Mall Gazette* does not list Henry Adams during the "season" alluded to. The reference may possibly be to another newspaper.

8. See Preface to the eleventh edition, January 15, 1872, p. vi.

9. Thwaites, *Outline of Glacial Geology,* p. 105.

10. Lyell had of course pointed out that the question of first causes was outside the sphere of geology and in his earlier *Antiquity of Man* (p. 421), with which Adams was familiar, he had indicated that such questions were irrelevant to scientific inquiry.

11. Gray, *Darwiniana,* p. 176.

12. *Nation,* VII (October 22, 1868), 335.

CHAPTER SIX: A YOUNG REFORMER AT THIRTY
1868-1870

A Power in the Land

1. Coleman, *Election of 1868,* p. 383.

2. Quoted in the New York *World,* March 16, 1867; *A Cycle,* II, p. 238; Ware, *Political Opinion,* p. 51. When on March 6, 1868, Adams went to Court in his uniform, Moran wrote in his Diary, "This is in opposition to law, but as he is going out of office he don't care. He hates the Senate and has no respect for the rights of his Secretary."

3. C. F. Adams, Jr., *Dana,* II, 331.

4. Coleman, pp. 281, 290.

5. At Charleston, October 16, 1868. Coleman, p. 330. He delivered only one speech at Charleston. Coleman erroneously lists two. On October 9 he was at Charlotte and not Charleston. New York *World,* October 10, 1868; quotation from the *Post* in the Boston *Transcript,* October 20, 1868; *Diary of Gideon Welles,* III, 488; New York *Times,* April 24, 1869.

6. Adams's recollection that he voted for Grant (*Education*, p. 260) seems at fault. Apparently he did not vote at all in the election. He was continuously in Washington after October 22; and his letter from there on November 5 (*Letters*, I, 148) refers to the election as just past. Absentee voting was not authorized in Massachusetts until 1917 when the state constitution was amended for the purpose. See article 45. *Letters* I, 147. See the curious recollection of the interview with Johnson in *The Education*, p. 245.

7. Rhodes, *History*, VI, 232; Welles, *Diary*, III, 486; Adams to Wells, *New England Quarterly*, XI (March 1938), 146-152; letter to Henry Reeve, listed among letters sent in Adams's 1868 *Date Book*. This list covers the period from November 1, 1868, to March 21, 1869, and tabulates fifty-seven letters, half of them to various members of his immediate family.

8. As "The Argument in the Legal Tender Case," signed H. B. A. Cf. the alignment of counsel in *The Education*, p. 249.

9. Howe, *Storey*, pp. 129, 124.

10. Henry Adams to Garfield, January 22, 1869 (Garfield Papers, Library of Congress).

11. For a full list of the "reform" writings of the period see L. A. Rose, "A Bibliographical Survey of Economic and Political Writings 1865-1900," *American Literature*, Vol. 15, January 1944, pp. 381-400.

12. *Republican*, May 1, 1869; the list of officers published in the Boston *Transcript*, April 22, 1869, does not contain the name of Henry Adams.

13. In another connection Henry Adams had also been advertised in Congress in a small way. Young Senator Sprague of Rhode Island, having attracted severe criticism for his attacks on predatory business interests, defended his character by reading into the record several hundred letters from public men as proof of his acumen and integrity. The first letter dated February 25, 1869, was from Adams requesting information about the rates of profit and the effect of currency manipulations on private interests. *Globe*, 41st Congress, 1st Session, 745. Mentioned in Boston *Transcript*, April 24, 1869.

Fundamental Principles

1. Thayer, *Hay*, I, 251. The date attributed—February 1867—is clearly a misprint. Adams was in London at that time.

2. Smith, *Garfield*, II, 758, 759.

3. *Republican*, November 27, 1869; *Nation*, IX (November 11), 415.

4. Adams to Cox, November 8, 1869, *Henry Adams and His Friends*, p. 43.

5. Harte's "Luck of Roaring Camp" was published in the *Overland Monthly* for August 1868; his "Outcasts of Poker Flat" followed in January 1869, the two stories giving great impetus to the vogue of local color in fiction. Mark Twain was already widely known for his "Jumping Frog of Calaveras County," which inaugurated a new era of humor on its appearance in 1867. He followed this *saut d'esprit* with *Innocents Abroad* in 1869. William Dean Howells with his *Venetian Life* (1866) and *Italian Journey* (1867) likewise promised better things to come. Young Henry James had already begun to make his mark as a reviewer for the *North American* and a contributor to *Galaxy* and the *Atlantic*. Perhaps one ought not to quarrel with Adams for ignoring, in addition, Bayard Taylor and E. C. Stedman, or such lesser, though more authentic, talents like John William De Forest, the author of *Miss Ravenel's Conversion* (1866).

6. *Letters*, I, November 23, 1869, "I am writing, writing, writing. You must take the New York *Nation* if you want to read me." March 7, 1870, "At the same time I write about two articles a month in the *Nation*. . . ." Cf. Springfield *Republican*, May 1, 1869, which stated that Henry Adams wrote articles for the *Nation* which "nobody, however, suspected as his." See also "The Argument in the Legal Tender Case." Another initialed article of his was published May 12, 1870, "The *North American Review* and the Hon. Elbridge G. Spaulding," attributed elsewhere in the same issue (p. 308) to Adams. By evident oversight his name was omitted from Pollak, *Fifty Years of American Idealism:* The New York Nation 1865-1915 and from "The 'Nation' and Its Contributors," *Nation*, Vol. 100, July 8, 1915. Adams seems to have been determined to bury this phase of his writing. He omitted all mention of the *Nation* articles in the list which he supplied for Justin Winsor's *Bibliographical Contributions*, 1881.

7. New York *Evening Post*, February 2, 1870, "A Delicate Suggestion"; *Globe*, 41st Congress, 2d Session, Pt. 2, p. 1113.

8. December 30, 1869. Ferleger, *Wells*, p. 278.

9. Warren, *Supreme Court in United States History*, III, 234; Nevins, *Fish*, pp. 305,.306; Sidney Ratner, "Was the Supreme Court Packed by President Grant?" *Political Science Quarterly*, Vol. 50 (September 1935). Adams did not resolve the contradiction between his desire to strengthen the Supreme Court and his even more intense desire to strengthen the Executive in its struggle with the Senate over the use of the appointing power. The Senate, in this

instance, seems to have been a greater champion of an independent court. In *Knox* vs. *Lee,* 12 Wallace, 457 (1871) the court reversed itself and held that the issuance of greenbacks had been constitutional, the result aimed at by President Grant.

10. *Nation,* X (May 12, 1870), 302.

11. *Nation,* X (May 12, 1870), 308; *ibid.* (May 26, 1870), 334.

Fame and Its Aftermath

1. Adams to Garfield, December 30, 1869 (Garfield Papers, Library of Congress). John Bonner, who figured importantly in Adams's article as a victim of the gold conspirators, anticipated Adams's exposé by writing one of his own for the April 1870 issue of *Harper's,* under the title "The Great Gold Conspiracy." He wrote without the benefit of Garfield's report.

2. Among the delegates to the "breakfast" were E. L. Godkin, David A. Wells, General Hawley, Horace White of the Chicago *Tribune,* Francis A. Walker, Charles Nordhoff of the New York *Post,* and Hugh McCulloch, ex-Secretary of the Treasury. Ogden, *Godkin,* II, 40; *Nation,* X (April 28, 1870), 263; Springfield *Republican,* April 26, 1870; Adams to Garfield, May 12, 1870 (Garfield Papers).

3. H. Davis, "From the Diaries of a Diplomat—James S. Pike," *New England Quarterly,* XIV, 108.

4. For a brilliant account of the various campaigns in the Wars of Erie, see Matthew Josephson, *The Robber Barons.*

5. See Smith, *Garfield,* p. 449.

6. Charles Francis Adams, Jr., "A Chapter of Erie," reprinted in Hicks, ed., *High Finance in the Sixties,* p. 31. This work also contains a reprint of "The New York Gold Conspiracy" by Henry Adams.

7. Tansill, *The United States and Santo Domingo,* declares (p. 341) that "even so shrewd an observer as Henry Adams was misled" into exaggerating the importance of Sumner's role in the fight against the Dominican treaty.

8. *Nation,* August 11, 1870; see Appendix, 1872, for the campaign pamphlet; *Wisconsin State Journal,* October 7, 1870, reprinted as a campaign broadside, *Political History. The Republican Party Defended. A Reviewer Reviewed.*

9. *Letters,* I, 185; Adams to Eliot, July 3, 1870, *Henry Adams and His Friends,* p. 45.

10. Gurney to Eliot, August 20, 1870 (Eliot Papers); Gurney to Godkin, August 27, 1870; Lowell to Godkin, August 23, 1870. All at Harvard College Library.

11. His salary was officially listed as $1600 and $400 for rooms.

College Records, XI, Harvard Archives; another entry gives $1700 and $300. *Ibid.*, p. 259. See also pp. 243, 334.

12. Gurney to Eliot, August 20, 1870. Charles Francis Adams served as an Overseer from 1869 to 1881. He was President of the Board from 1874 to 1878. When it became evident that he would not enter Grant's cabinet, the Board of Overseers offered him the presidency of Harvard. C. F. Adams, Jr., *C. F. Adams*, p. 379. Adams declined the honor and it thereafter was tendered to Eliot. James, *Eliot*, I, 191.

CHAPTER SEVEN: HARVARD COLLEGE ONCE MORE 1870-1877

The Academic Rebel

1. Adams to Norton, January 13, 1871, *Henry Adams and His Friends*, p. 54.

2. Adams to Cox, November 17, 1870, *Henry Adams and His Friends*, p. 48. Adams to Norton, January 13, 1871. On the Harvard "revolution" see, for example [John Fiske], "The Presidency of Harvard College," *Nation*, December 31, 1868; C. W. Eliot, "The New Education," *Atlantic Monthly*, 1869, XXIII; Noah Porter, "The American College and the American Public," *New Englander*, 1869, XXVIII; Charles A. Bristed, "The Dispute about Liberal Education," *Lippincott*, 1868, II; John Fiske, *Letters*, p. 180.

3. Laughlin, "Some Recollections," p. 579. Adams had a liking for "unusual and tentative explanations of puzzling problems." Adams to Maine, February 22, 1875, *Henry Adams and His Friends*, p. 63; Lindsay Swift, "A Course in History at Harvard College in the Seventies"; Adams, Class Record Book 1873-1877 (MS); *Johns Hopkins University Studies in History and Political Science*, vol. 2 (1884), p. 88; Herbert B. Adams gives specimen examinations in Adams's courses in his *Study of History in American Colleges and Universities* (Washington, 1887); see also Ephraim Emerton, "The Practical Method in Higher Historical Instruction" in *Heath's Pedagogical Library*, edited by Stanley Hall (1884).

4. *Letters*, I, 217, October 23, 1871. The number of courses which Adams taught in his first year, 1870-71, is not clear. *Letters*, I, 194, speaks of "nine hours a week in the lecture room," implying three courses, normally of three hours each. The "Final Returns" show but two courses for 1870-71:

> History I [The General History of Europe from 987]. Textbook: Duruy's *Moyen Age*. 13 juniors.
>
> History II [Medieval History]. Textbook: Hallam's *Middle Ages*. 45 seniors, 42 juniors.

The pamphlet "Studies of Seniors, Juniors, and Sophomores in the years 1870-71 and 1871-72" announces History II, for seniors, as "History of the Period from the end of the Middle Ages to 1648"; whereas History II, for juniors, is described as "Medieval History." It would appear, therefore, that although three courses had originally been scheduled—History I; History II, for juniors; History II, for seniors—one course was dropped, probably History II (for seniors), and History II (for juniors) given to seniors and juniors together. It is equally possible that the class met in two sections: one for seniors and one for juniors. Letter from Clifford K. Shipton, Custodian of the Harvard University Archives.

Courses taught by Adams after the first year were all three-hour electives:

1871-1872

History of Germany, France, and the Church (from the Eighth to the Fifteenth centuries). 33 juniors.
Medieval Institutions ("for Honors"). 7 juniors.
History of England (to the Seventeenth Century). 15 seniors.

(Out of residence 1872-1873)

1873-1874

II General History of Europe—Tenth to the Sixteenth Century. 3 seniors, 65 juniors.
III Medieval Institutions ("only for candidates for honors"). Two theses required. 5 seniors, 7 juniors (plus 3 extra students).
IV History of England to the Seventeenth Century (Constitutional and Legal). 20 seniors.

1874-1875

(Shortly before turning over History II to Ernest Young, one of his doctoral candidates, Adams prepared a six-page printed syllabus for the use of the students, entitled "Political History of Europe from the 10th to the 15th Century." The title suggests that Adams, like the Englishman Freeman, tended to regard history as "past politics." The Syllabus lists 298 items like the following: "156. The contest about the Papacy in 1046," "247. The Battle of Muret; when and where did Simon of Montfort die?")

III Medieval Institutions (advanced course). 13 juniors and seniors.
IV History of England to the Seventeenth Century. 17 seniors.
V Colonial History of America to 1789. 27 juniors and seniors.

Graduate seminar in Anglo-Saxon law for Ph.D. candidates. (Authorized by the President and Fellows on June 8, 1874.)

1875-1876

III Medieval Institutions (advanced course). 16 juniors and seniors.

IV History of England to the Seventeenth Century. 13 upperclassmen.

V Colonial History of America to 1789. "Ability to read German will be of great importance, especially in Course 3." 16 upperclassmen.

Graduate course: Early English Institutions.

1876-1877

III Medieval Institutions (advanced course). 11 upperclassmen.

IV History of England to the Seventeenth Century. 23 upperclassmen.

VI History of the United States from 1789-1840. 57 upperclassmen.

Graduate course: Medieval Institutions.

(Annual Catalogues of Harvard University; Manuscript Class Record Book of Henry Adams 1873-1877; Harvard College Record XI, XII; *Letters*, I, 265.)

5. Swift, "A Course in History . . ."

6. December 14, 1875, *Henry Adams and His Friends*, pp. 73-74; Morison, "Edward Channing"; Swift, "A Course in History"; Laughlin, "Some Recollections."

7. Edward D. Bettens. Printed letter to Taylor.

8. Emerton in Morison, *Three Centuries;* Taylor, *Human Values*, p. 40; Taylor, "The Education," *Atlantic*, vol. 122, p. 490; Thwing, *Guides*, p. 231.

9. Brooks Adams, "Heritage of Henry Adams," p. 6; Morse, *Perry*, p. 63; Howe, *Storey*, p. 241; Bent, *Holmes*, p. 65; R. B. Perry, *James*, I, 109, 360; Mrs. Henry Adams, *Letters*, p. xv.

From Tea Kettle to Steam Engine

1. Mott, "One Hundred and Twenty Years," p. 164; Mott, *American Magazines*, pp. 31, 32; Adams to Norton, January 13, 1871, *Henry Adams and His Friends*, p. 54.

2. Adams to Cox, November 28, 1870, *Henry Adams and His Friends*, p. 45.

3. Adams to Wells, *New England Quarterly*, XI, 148; Adams to Cox, November 17, 1870, *Henry Adams and His Friends*, p. 48.

4. Tilden, *Letters*, I, 288; Henry Adams, *Letters*, I, 197; *Nation*, XII (January 26, 1871), 62; Adams to Norton, January 13, 1871, *Henry Adams and His Friends*, p. 53. See New York *Times*, January 10, 1871, for praise of C. F. Adams's railroad article.

5. Mott, "One Hundred and Twenty Years," p. 164.

6. Moran Diary, October 18, 1867. "Mr. Adams has been speaking in very disrespectful terms of Mr. McHenry and the Atlantic and Great Western Railway. He hates that gentleman and has joined the ["Boston"?] chorus of Sturgis, Lampson, and Morgan to hunt him down." Moran thought McHenry "a public benefactor."

7. Adams to Wells, January 17, 1871, in "Henry Adams as Editor"; *Letters*, I, 207-210; Adams to Cox, September 30, 1871, Cox Papers.

8. W. S. Robinson ("Warrington") wrote Sumner that the nomination of Charles Francis Adams would be "a great mistake, for he and his family represent too much the anti-popular element—the sneering, sniffling element, which can never have a permanent success in our politics." Quoted in Ross, *Liberal Republican Movement*, p. 85.

Editorial Principles

1. Mott, "One Hundred and Twenty Years."

2. *Unpublished Letters of Bayard Taylor*, p. 153; proof sheets in the Harvard Library.

3. See above, p. 184, for his fling at his brother Charles.

4. Adams's extravagantly high opinion of Sohm's work was probably the result of his acquaintance with Marcel Thévenin, whose translation of the *Lex Salica* was the subject of Adams's review. Thévenin, said Adams, had "assumed the task of spreading the influence of Sohm's investigations beyond Germany" and he had in this work rendered "the terrible periods and German archaisms of Professor Sohm into intelligible French." Another affiliation appears to have been Dr. Leopoldo Yanguas, who sent a presentation copy of his *Estudio sobre el Valor de las hetras Arabigas en el Alfabeto Castellano* (Madrid, 1874). A more important affiliation was Gabriel Monod, President of the Historical Section of the École des Hautes Études (founded in 1868) and co-author with Charles Bémont of *Medieval Europe 395-1270* (translated by Mary Sloan). Adams conferred with him in 1879. See Adams to Godkin, December 25, 1879, *Henry Adams and His Friends*, p. 95.

5. Adams to Maine, February 22, 1875. *Henry Adams and His Friends*, p. 63; Adams to Bancroft, March 25, 1875, *Henry Adams and His Friends*, p. 65.

6. The comments in this and the preceding paragraph in the text are taken from his reviews of the works of these men.

Cargill, in his "Medievalism of Henry Adams," explains Adams's frequent attacks on Freeman as motivated by his personal partiality for the elder Palgrave. This explanation does not seem adequate. As Sir Francis Palgrave died early in 1861, Adams never became acquainted with him. He did not become friendly with Francis Palgrave, the son of the historian, until after he met young Palgrave's brother-in-law, Charles Milnes Gaskell, in 1863. The younger Palgrave was on the very best terms with Freeman and apparently did not at all resent Freeman's severe criticism of his father's literary style in a review of Palgrave's *History of Normandy* for the *Edinburgh Review* in 1859 (vol. 109). In 1862 Palgrave, the younger, sent two posthumous volumes of his father's book to Freeman for criticism. Freeman enthusiastically urged publication (Freeman, *Letters*, I, 322, 323) and in his review of the work (*Edinburgh,* January 1865) he praised it highly while acknowledging its serious imperfections of style. Throughout his life Freeman always spoke of Palgrave in terms of admiration and respect. Cf. Stephens, *Freeman,* I, 116. Adams himself in a review of Stubbs in the *North American* (vol. 119, p. 235) wrote that "the excellent books of Kemble and Palgrave were becoming antiquated." If Adams and his students adopted a "Palgravian Thesis" it must have been wholly through inadvertence. The list of "Titles of Works Cited" in Adams, and others, *Essays in Anglo-Saxon Law* does not mention Sir Francis Palgrave. Adams may well have been aware of Freeman's serious shortcomings both of scholarship and temperament. Cf. G. P. Gooch, *History and Historians of the Nineteenth Century,* p. 349; J. R. Green, *Letters,* p. 226 and *passim.*

In his first review of Freeman Adams thought he had "caught him out very cleverly." He wrote: "Mr. Freeman stands in the very front rank of living English historians. He is the legitimate successor of Hallam, Palgrave, and Grote. Having said so much, we have said all that is required in recommendation of this book, the contents of which are rather necessary to an elementary education than to the attainment of any very advanced knowledge . . . As usual with his controversial work, he ends in producing a feeling of reaction against himself and his very just though rather commonplace ideas . . . Barring Mr. Freeman's most inveterate prejudices, he is . . . a hard student and an honest workman . . . He has read the great German historians, and he probably admires them, but he has certainly failed to understand either their method or their aims . . .

"In spite of his labors the history of the Norman Conquest and a statement of Anglo-Saxon institutions still remain as far from realization as ever."

This review found a reader in Freeman himself, who, as Adams put it, did "condescend to write a very droll reply." Whatever the drolleries, Adams attempted a sort of *amende honorable* two years later in a notice of a new issue of *The History of the Norman Conquest,* which he thought "calculated to improve his temper": "A new edition of Mr. Freeman's great work is a gift to the student of English history which he ought to value highly . . . perhaps at some future day, when the work is completed, this *Review* will be obliged to discuss it as carefully as its great merits deserve [Perhaps an echo of the *Quarterly Review* (1873), vol. 135, p. 176.] . . . For carefulness in study this work is without a rival among English histories . . . His patriotic enthusiasm for his Saxon ancestors . . . is an element of the book which has positive value, if only because it is a healthy reaction against the old tendency to consider everything good in civilization as due to Rome and Greece." The remainder of the review deprecated, however, Freeman's interpretation of the compurgation of one of the principal actors in a celebrated Anglo-Saxon homicide. "Law is indeed Mr. Freeman's weakest point." This last observation is oddly revealing when we come upon his own confession to Lodge (*Letters,* I, 291) concerning Lodge's doctoral dissertation that "above all I cannot understand law." A case might be made out that Adams was deeply indebted to Freeman for many of his ideas about English history and the writing of history and that his criticism was a form of compensation. His almost savage annotations of Edmund Burke's writings suggest the pattern of his habitual reactions. The curious student may pursue the many striking parallels between the positions taken up by Adams and his scholars and those asserted by Freeman in his *History of the Norman Conquest.* Cf., for instance, Freeman, *Norman Conquest* (1873), I, 62, 63. (See vol. V, p. 578, for praise of Palgrave's account of the origin of the "feudal relation.") Cf. also Freeman's Rede lecture, 1872, "The Unity of History," for interesting parallels in general ideas. It is also likely that Adams picked up some of his opinions, favorable and adverse, from the English reviews. Freeman's good friend J. R. Green was fairly frank in his appraisal of the work of his master. Cf. *Saturday Review,* April 13, 1867, and his succeeding reviews of the *Norman Conquest;* also Green's *Historical Studies.*

7. Green, *Letters,* p. 408.

8. Lodge, *Early Memories,* pp. 244, 245. Cf. *Letters,* I, letters to Lodge.

Wedding Journey of a Scholar

1. *Letters,* I, 223.

2. Howells, *Life and Letters,* p. 172. "The October issue was left *planted,* by both the late editors, and I'm putting it together

from such material as I can get, and proofreading it"; Adams to Reid, May 15, 1872, *Henry Adams and His Friends*, p. 56. A brief notice of the wedding appeared in the Boston *Advertiser*, June 29, 1872; Mrs. Henry Adams, *Letters*, p. 14. Cf. Lowell, *Letters*, II, 80, ed. Norton; Mrs. Henry Adams, *Letters*, pp. 65, 477-78.

3. Adams to Godkin, December 16, 1885, *Henry Adams and His Friends*, p. 158.

4. Mrs. Henry Adams, *Letters*, for this and many other details of their journey.

5. An account of his career and attainments is given by Frederick Tuckerman in American Economic Association Publications, series 3, vol. 7 (1906), p. 474. See also *Nation*, November 2, 1871, review of *North American* for October.

6. Lowell, *New Letters*, pp. 198-201.

Work and Plans

1. *Letters*, I, 256. As his own Class Record book indicates only three three-hour courses, he may possibly have broken the very large class in History II into two sections; or he may have divided History III, Medieval Institutions, into two seminars, one for undergraduates and one for graduates, preparatory to formally instituting the graduate seminar in the following year.

2. Lodge, *Early Memories*, p. 240.

3. Adams to Parkman, September 26, 1874; to Bancroft, March 25, 1875, *Henry Adams and His Friends*, p. 65; Emerson, *Letters*.

4. Cushing, *Index of the North American Review*. Unfortunately the *Index* is not wholly accurate. As a rule the critical notices were anonymous. Thirteen, however, bore the initials of their authors.

5. See above, p. 212, note 4. It is possible that the original inspiration for his undergraduate honors class and for his graduate class came from his acquaintance with Professor Charles Kendall Adams, who established a seminar in history at the University of Michigan in 1869. C. K. Adams had studied in German universities and appears to have been the American pioneer of the historical seminar. Cf. last three references, note 3, p. 212. He visited Harvard in March of 1871 (Eliot Papers, 6) and evidently sought out Henry Adams, for he subsequently contributed five articles to the *North American*, beginning with the October 1871 issue.

6. Harvard College Record XII, 127; Adams to Gaskell, May 24, 1875. *Henry Adams and His Friends*, p. 66.

7. Adams to Eliot, April 4, 1876. *Henry Adams and His Friends*, pp. 76-77. Adams's intensive reading for the course was characteristic of his method. To check his hypothesis of the common origin of the American settlements he enlisted the expert help of one of Dr. Palfrey's friends, the New England historian Samuel Foster

Haven. Adams to Haven, November 23, December 7, 1874. *Henry Adams and His Friends,* pp. 60-62. Wherever possible he went to the original sources: *Letters,* I, 298, "I am still buried in avalanches of State Papers"; *ibid.,* "I have read all Timothy's [Pickering] diplomatic papers lately"; *Proceedings,* Massachusetts Historical Society, 1876-77, December 1876, he was given permission to copy from the voluminous Heath Papers; he purchased a set of the *Niles Register; Letters,* I, 293, "I have read Marbury *vs.* Madison with care, and talked with Wendell Holmes on the theory of the constitution."

Essays in Anglo-Saxon Law

1. Quotations are from his reviews of the works of these men.

2. Adams to Maine, February 22, 1875. *Henry Adams and His Friends,* p. 64; Adams to Morgan, April 29, 1876. *Henry Adams and His Friends,* p. 78; Stern, *Morgan,* p. 119.

3. *The Education,* p. 368.

4. Cf. the earlier statement of his position in his review of John R. Green's *Short History of the English People* (July, 1875): "The one great constitutional machine which characterized all Teutonic society in its earliest historic phase was, therefore, the Hundred or shire court, from which was developed all the public and private law of the Anglo-Saxon time, army, king, witan, and the whole administration of justice, even to the point of absorbing much church law; and this same primeval institution, surviving shock of conquest, civil war, and social decay, even after the steady drain of centuries which carried its powers one by one into other hands, still retained force enough in the thirteenth century to become the foundation of Parliament."

5. In the Preface to the 1877 edition of his *Norman Conquest,* Freeman, harking back to Adams's earlier attack, remarked that his anonymous critic was evidently unaware that there were a number of versions of the Anglo-Saxon Chronicle. Freeman's gibe must be taken as a piece of satirical criticism as the existence of various manuscripts of the Chronicle was attested by scholars like John M. Kemble whose *The Saxons in England* (London, 1849), which alludes to the manuscripts (for example, II, 218, 219, references to Thorpe—224, 236), was well known to Adams and his scholars. Stubbs, *Constitutional History* (Oxford, 1875-1884), pp. 86, 92, 96, 125; *Letters,* I, 323.

6. *Law Magazine and Review,* III (1877-78), pp. 120, 121; *American Law Review,* II (1876-77), pp. 327-31; *North American Review,* vol. 124 (1877), p. 329; *Johns Hopkins Studies in Historical and Political Science,* II (1884), 6, 88. For an account of the in-

troduction of American institutional history at Johns Hopkins University, see *Circular of Information*, Bureau of Education (Washington, 1887). Henry Adams's part in this development is told by Herbert B. Adams in "New Methods of Study of History," *Johns Hopkins Studies in Historical and Political Science*, vol. 2 (1884), p. 101. The continuity of the Germanic village community in New England had been originally suggested to Sir Henry Maine by an article in the *Nation*, communicated by Prof. W. F. Allen of the University of Wisconsin. "It was determined as early as 1877, after consultation with Professor Henry Adams . . . to apply this principle of continuity to the town institutions of New England."

7. Charles Gross, *Sources and Literature of English History*, 2d ed. (London, 1915), p. 295; Percy Winfield, *Chief Sources of English Legal History* (Cambridge, 1925), p. 53. Charles A. Beard, "The Teutonic Origins of Representative Government," *American Political Science Review*, vol. 26 (1932), p. 44. He thinks it unlikely that the shire and hundred courts were democratic assemblies of free delegates: "Were they stalwart freemen going to the assemblies to stand for their rights and those of their constituents. Perhaps not." J. E. A. Jolliffe in "The Origin of the Hundred in Kent," *Historical Essays in Honour of James Tait*, edited by J. Edwards (Manchester, 1933), declares that "the most difficult problems of early institutional history still centre about the hundred." Similarly, a reviewer in the *American Historical Review*, April 1945, alludes to the "favorite old enigma of the hundred." An excellent early summary of the literature of the subject is supplied by Paul Vinogradoff in the Preface to his *Villainage in England* (Oxford, 1892). He writes: "For the representatives of the New School this 'original Teutonic freedom' has entirely lost its significance, and they regard the process of social development as starting with the domination of the few and the serfdom of the many." The best introduction to the immense complexities of the questions raised by Adams and his students is Henry J. Ford, *Representative Government*, especially Chapters I-VIII. For a recent discussion of the whole subject see F. M. Stenton, *Anglo-Saxon England* (Oxford, 1943). The curious persistence of the Teutonic theory may be seen in one of the great standard texts, W. S. Holdsworth's twelve-volume *A History of English Law*. Cf. vol. II (1923), pp. 12-14; and also in F. Maitland and F. Montague, *A Sketch of English Legal History* (New York, 1915), pp. 17, 21. "The desire to prove some racial, social, or constitutional thesis generally leads to attempts to prove it by historical precedents." Holdsworth, II, 18, quoting from Reeves, *History of English Legal Institutions*. Cf. also Hugh M. Clokie, *Origin and Nature of Constitutional Government* (London, 1936), pp. 13-15.

8. Cf. E. A. Freeman, *Growth of the English Constitution* (London, 1872), p. 10, "sovereign power is vested in the whole people." See also pp. 13, 14, 18-21. As to kings, the "nation chose them and the nation could depose them." *Ibid.*, p. 30.

9. For a provocative analysis of those contradictions and of the limitations of Adams's work, see Cargill, "The Medievalism of Henry Adams." See above, p. 232 and note 6.

The Primitive Rights of Women

1. Adams to Morgan, May 14, 1875, "Mr. Hubert Bancroft's second volume on the Indians of the Pacific Coast requires a notice of a few pages." Morgan Collection, University of Rochester. Adams to Morgan, October 3, 1875: "The subjects he treats deserve scientific analysis, and it will be a disgrace to let such work go out as the measure of our national scholarship." Adams to Morgan, June 3, 1876. The last two quotations in this paragraph and the succeeding quotations from Adams's letters to Morgan were first given in Stern, *Morgan.* Cater, in *Henry Adams and His Friends,* gives the entire texts of these letters, but omits the May 14 letter with its clue to the genesis of Morgan's first article, and misdates Adams's inquiry concerning the marriage customs of the American Indians. The correct date is June 3, 1876 and *not* 1878. It was written *before* Adams's Lowell lecture, "Primitive Rights of Women."

2. On the stand of John Quincy Adams (II) see Ware, *Political Opinion,* p. 188. A book which may have interested Adams as a one-time traveler in Italy, H. Taine's *Italy: Rome and Naples,* 4th ed. (New York, 1873), contains a number of suggestive parallels to his thought. In the translation issued in 1876 there is mention (p. 336) of the recent growth of the interest in the Virgin ". . . the all-powerful mediatress of the human species . . ." This interest seems to have accompanied the issuance of the papal bull "Ineffabilis Deus," December 8, 1854, which made the doctrine of the immaculate conception an article of faith. The book contains a number of analogies to the laws of physics; for example, Taine says (p. 336) of the archaic rites of the Church, "All these are dead forces, that is to say, due to an acquired momentum, and which act only through the natural inertia of human matter."

3. Several of the many learned annotations in Adams's desk copy of Merkel's *Lex Salica* concern the nature of the marriage relation under early law. (Western Reserve University Library.) In this general connection mention ought also to be made of Comte's discussion of the importance of woman and the family and of Buckle's celebrated lecture on woman.

4. For his conception of instincts innate in "human nature" see also his *History*, VI, 127. Cf. Brooks Adams, *Degradation*, p. viii.

5. On this point see also Brooks Adams, *Degradation*, pp. 118 ff.

6. George Elliott Howard, *History of Matrimonial Institutions* (Chicago, 1904), p. 9 and *passim;* and by the same author *The Family and Marriage* (Lincoln, Nebr., 1915).

7. The reviewer for the Boston *Journal* and the Boston *Transcript* (December 11, 1876) attempted only brief summaries of Adams's lecture. As the language is largely identical one may assume they were written by the same person. The summaries are much too sketchy to indicate whether the lecture as finally printed in Adams's *Historical Essays* differed substantially from the lost original.

American History and the Constitution

1. Adams to Bancroft, January 10, 1884, *Henry Adams and His Friends,* p. 129, explains Adams's share. Compare to the opinions of John Quincy Adams expressed in two letters to George Bancroft (October 25, 1835, and March 31, 1838). John Quincy defended government by an élite against mere democracy of persons. He could not see the rise of a Christian philosophical "humanized democracy" in New England which Bancroft did. The democracy of Thomas Paine and others was a "Government for wild beasts and not for men." New York Public Library, *Bulletin*, X (1906). Charles Francis Adams in *Works of John Adams*, I, 363.

2. Laughlin, "Some Recollections." *Letters*, I, 295, Adams to Lodge, "I admire them [the Puritans] as much as ever, but I shall not deny that they were intolerant even according to the age in which they lived."

3. Cf. Henry Cabot Lodge, *A Short History of the English Colonies in America* (New York, 1881) (dedicated to Henry Adams "in token of gratitude and friendship"), pp. 519, 520.

4. *Nation,* January 3, 1878.

5. Adams to Eliot, March 12, 1877, *Henry Adams and His Friends,* p. 80.

Political Reformer: Final Phase

1. Adams to Gilman, November 17, 1875. Holt, "Henry Adams and Johns Hopkins University," *New England Quarterly,* XI (1938), 633.

2. Quoted by James T. Bixby, in a review of Taine's *On Intelligence* (1872) in the *North American Review,* October 1873, p. 402; *Letters*, I, 305.

The interest which Henry Adams displayed in the scientific investigation and measurement of society was very much a reflection of his time. His college classmate George E. Pond wrote in the *Galaxy* for May 1873: "The Taine of the Twentieth Century who shall study the literature of the Nineteenth will note an epochal earmark. He will discover a universal drenching of *belles lettres* with science and sociology, while the ultimate dominant tinge in our era he will observe to be Darwinism. Not only does all physical research take color from the new theory, but the doctrine sends its pervasive lines through poetry, novels, and history." Quoted in Mott, *American Magazines,* III, 106. Mott also points out that Tyndall's lectures on physics in the winter of 1872-73 in New York were enormously popular. Freeman in his lecture on *The Unity of History* (London, 1872), saw the eternal character of Rome as a main clue to the unity of history and may have been one of those who inspired Adams to imitate Gibbon's meditations on the steps of the Ara Coeli.

The notion that progress was a function of the human mind was discussed by Walter Bagehot in *Physics and Politics* (London, 1873), a work which Adams may have turned to after Gamaliel Bradford's enthusiastic praise of Bagehot's *English Constitution* in the January 1874 *North American.* Cf. also Martineau's article in the *Contemporary Review,* April 1872, "The Place of Mind in Nature, and Intuition in Man." In addition to Buckle, Mill, and Comte, Adams doubtless read during his study years in England John W. Draper's *History of the Intellectual Development of Europe* (New York, 1863) and William Lecky's *History of Rationalism* (London, 1865). In a *North American* article called "The Laws of History" (July, 1869), John Fiske declared that "in the science of history, the deductive method must be used, no less than in astronomy . . ." He cautioned, however, that "the law of universal evolution can no longer supply the precise kind of information we desire . . . Though it is the ultimate law of history, it is silent respecting the differential characteristics by which a historic event is distinguished from a physical one." These cautions, as we know, Adams ultimately disregarded when he deduced his "Dynamic Theory of History." Another highly suggestive article was Frederic Hedge's "The Method of History" in the October 1870 *North American.* Hedge postulated a theory of periodicity by which the "course of history becomes a *spiral* movement at once revolutionary and progressive . . . whose action is calculable," an idea reinforcing Adams's early conception of tides in human affairs, which had been excited by the experiments with the Armstrong rifles. Cf. *History of the United States,* V, 361; VI, 123, 436.

3. Kraus, *A History of American History,* p. 334, points out that "throughout his volumes [*History of the United States*] Adams uses

the terminology of the physicist, and often with marked effect." Zucker, *Philosophy of American History*, I, 52, calls attention to Adams's use of the Second Law of Thermodynamics in the concluding chapter of the *History*, that is, in the concept of a democratic ocean in a state of equilibrium. Cf. *History of the United States*, I, 176, "Of all historical problems the nature of a national character is the most difficult and the most important."

4. *Nation*, February 17, 1876, p. 118.

5. Adams to Garfield, December 13, 1873, Garfield Papers, Library of Congress.

6. Adams to Wells, April 16, 1875: "Henry Adams as Editor," *New England Quarterly*, March 1938.

7. Announced in his letter to Charles M. Gaskell of May 24, 1875, *Henry Adams and His Friends*, p. 65.

8. A striking portrait of Boston and its mind appears in Adams's review of Ticknor, *North American Review*, vol. 123, July 1876.

9. Schurz to Adams, February 24, 1876, Massachusetts Historical Society.

10. Munroe, *Walker*, p. 159, Adams to Walker, February 29, 1876. His interest in a newspaper which would support the reform program of the Liberal Republican group did not die. In 1881 he put in $20,000 as a member of a syndicate which purchased the New York *Evening Post*, edited by Edwin L. Godkin. Adams to Godkin, September 26, 1881. *Henry Adams and His Friends*, p. 114.

11. Based on information supplied by James D. Phillips, vice-president of Houghton Mifflin Company, and on Adams's letters.

12. His plans for the article are mentioned in *Letters*, I, 287.

13. "Charles Francis Adams, Jr.," *Dictionary of American Biography*.

14. Lodge to Bancroft, October 10, 1876, Massachusetts Historical Society; *Nation*, October 12, 1876. In another article in the same issue a favorable commentary on "The Independents in the Canvass" erroneously attributes it to Charles Francis Adams, Jr., alone.

Emancipation

1. See above, p. 212, note 4 for official statistics.

2. C. F. Adams, Jr., in January 1868 *North American;* Adams to Godkin, August 6, 1881, *Henry Adams and His Friends*, p. 110; Adams to Walker, Monroe, *Walker*, p. 159; James B. Conant, *History and Traditions of Harvard College*, p. 29.

3. Adams to Gilman, October 26, 1876, Holt, *New England Quarterly*, XI, 635; Harvard College Records XII, 251.

4. Taylor, "The Education of Henry Adams," *Atlantic,* vol. 122, p. 490.

5. Adams, Washington Date Book, 1868, Massachusetts Historical Society.

6. Adams to Gaskell, May 24, 1875, *Henry Adams and His Friends,* p. 65.

7. *Holmes-Pollock Letters,* II, 18.

8. From "Characters," by Caroline Sturgis Tappan, quoted in *Yale Review,* Winter, 1943, p. 330.

9. Adams to Gilman, January 21, 1878, Holt, *New England Quarterly,* XI, p. 637.

SELECTED BIBLIOGRAPHY

(Starred items contain letters or parts of letters by Henry Adams)

BIOGRAPHIES, MEMOIRS, AND LETTERS

Adams, Brooks. "The Heritage of Henry Adams," *Degradation of the Democratic Dogma*. New York, 1919.

* *A Cycle of Adams Letters 1861-1865*. Boston, 1920.

Adams, Charles Francis (Sr. and Jr.). "Campaigning with Seward in September, 1860," *Minnesota History*, VIII, 1927.

Adams, Charles Francis, Jr. *An Autobiography*. Boston, 1916.

—— *Charles Francis Adams*. Boston, 1900.

—— *Richard Henry Dana*. Boston, 1890.

Adams, Henry. *The Education of Henry Adams* (privately printed, 1907). Boston, 1918.

* —— *Letters of Henry Adams, 1858-1891*. Edited by Worthington C. Ford. Boston, I, 1930; II, 1938.

* —— *Henry Adams and His Friends*. Compiled by Harold Dean Cater. Boston, 1947.

* —— "Seventeen Letters," edited by F. B. Luquiens. *Yale Review* (NS), X, October 1920.

* —— "Six Letters," edited by A. S. Cook. *Yale Review* (NS), X, October 1920.

* —— "Three Letters," edited by A. S. Cook. *Pacific Review*, II, September 1921.

* —— "Henry Adams as Editor: A Group of Unpublished Letters Written to David A. Wells," edited by John E. Alden. *New England Quarterly*, XI, March 1938.

* —— "Letters to Frewen," edited by Shane Leslie. *Yale Review*, vol. 24, 1934-35.

* —— "Unpublished Letter to Sir Henry Maine," edited by Harold J. Laski. *Nation*, vol. 151, August 3, 1940.

* —— "Henry Adams and the Johns Hopkins University," edited by W. S. Holt. *New England Quarterly*, XI, 632.

* —— "A Note on Henry Adams" (Letters to Charles F. Thwing, President of Western Reserve University). *Colophon*, V, Pt. 17.

* —— *Letters to a Niece and Prayer to the Virgin of Chartres, with a Niece's Memories.* Mabel La Farge. Boston, 1920.

* Adams, James Truslow. *The Adams Family.* Boston, 1931.

—— *Henry Adams.* New York, 1933.

Adams, John. *Works of John Adams.* Edited by Charles Francis Adams. Boston, 1850-1856.

Adams, John Quincy. *Memoirs of John Quincy Adams.* Edited by Charles Francis Adams. Philadelphia, 1874-1877.

—— *Writings of John Quincy Adams.* Edited by Worthington C. Ford. New York. 1913-1917.

* Adams, Marian Hooper. *Letters of Mrs. Henry Adams.* Boston, 1936.

* Agassiz, Alexander. *Letters and Recollections of Alexander Agassiz.* Boston, 1913.

Anderson, Larz. *Larz Anderson; Letters and Journals of a Diplomat.* Edited by Isabel Anderson. New York, 1940.

Anderson, Nicholas L. *Letters and Journals of General Nicholas Longworth Anderson.* Edited by Isabel Anderson. New York, 1942.

Barnes, T. W. *Memoir of Thurlow Weed.* Boston, 1884.

Barrows, Chester L. *William M. Evarts.* Chapel Hill, 1941.

Bartlett, J. Gardner. *Henry Adams of Somersetshire.* New York (privately printed), 1927.

Beatty, Richard C. *Bayard Taylor Laureate of the Gilded Age.* Norman, Okla., 1936.

Beer, Thomas. *Hanna.* New York, 1929.

* Belmont, Perry. *An American Democrat.* New York, 1940.

Bent, Silas. *Justice Oliver Wendell Holmes.* New York, 1932.

Bettens, Edward. Printed Letter to Thomas Fenton Taylor. "Two Great Teachers." (At Massachusetts Historical Society.)

Bright, John. *The Diaries of John Bright.* Edited by R. Walling. London, 1930.

Cecil, Lady Gwendolen. *Life of Robert Marquis of Salisbury.* London, 1921.

Chanler, Mrs. Winthrop. *Roman Spring.* Boston, 1934.

Conway, Moncure. *Autobiography, Memories and Experiences.* New York, 1904.

* Cortissoz, Royal. *John La Farge.* Boston, 1911.

* —— *Life of Whitelaw Reid.* New York, 1921.

Crawford, Mary. *Famous Families of Massachusetts.* Boston, 1930.

Crowninshield, Benjamin W. *A Private Journal 1856-1858.* Edited by Francis B. Crowninshield. Cambridge, 1941. Privately printed.

———— "Private Journal commenced at Hanover, October 20th, 1858." Unpublished manuscript owned by Francis B. Crowninshield.

Darwin, Francis, ed. *Life and Letters of Charles Darwin*. London, 1876.

Dennett, Tyler. *John Hay: From Poetry to Politics*. New York, 1933.

———— "Five of Hearts," *Scholastic*, March 14, 1936.

Destler, Chester McArthur. *American Radicalism 1865-1901*. New London, Conn., 1946.

Dodge, Augusta. *Gail Hamilton's Life and Letters*. Boston, 1901.

Elliott, Maud H. *Three Generations*. Boston, 1923.

Elsey, George M. "The First Education of Henry Adams," *New England Quarterly*, XIV, December, 1941.

Emerson, Ralph Waldo. *Letters of R. W. Emerson*. New York, 1939.

Everett, Edward. "Peter Chardon Brooks," *Lives of American Merchants*. Edited by Freeman Hunt. Cincinnati, 1858.

Ferleger, Herbert. *David A. Wells*. Columbia diss., New York, 1942.

Fiske, John. *Letters of John Fiske*. Edited by Ethel F. Fisk. New York, 1940.

Foner, Philip S. *History of the Labor Movement in the United States*. New York, 1947.

Frothingham, Paul. *Edward Everett*. Boston, 1925.

Fuess, Claude. *Carl Schurz Reformer*. New York, 1932.

Garrison, Wendell. *William Lloyd Garrison*. New York, 1885-1889.

Gilder, Richard W. *Letters of Richard Watson Gilder*. Edited by Rosamond Gilder. Boston, 1916.

Grant, Robert. *Fourscore*. Boston, 1934.

Grattan, C. Hartley. *The Three Jameses*. London, 1932.

Green, John Richard. *Letters of John Richard Green*. Edited by Leslie Stephen. New York, 1901.

Green, Samuel S. "Memoir of the Rev. Edward G. Porter. '58'," *Transactions*, 1900-1902, Colonial Society of Massachusetts, VII.

Griscom, Lloyd C. *Diplomatically Speaking*. Boston, 1940.

Gwyn, Stephen. *Letters and Friendships of Sir Cecil Spring-Rice*. London, 1929.

Harriman, Mrs. J. Borden. *From Pinafores to Politics*. New York, 1923.

Harrison, Frederic. *Autobiographic Memoirs*. London, 1911.

* Hay, John. Letters and Extracts from Diary. Washington, printed but not published, 1908. (Keys at Illinois Historical Society, Springfield; University of Chicago Lincoln Library; Massachusetts Historical Society.)

Hayes, Rutherford B. *Diary and Letters of Rutherford Birchard Hayes*. Edited by Charles Williams. Columbus, Ohio, 1922-1926.

Hesseltine, William R. *Ulysses S. Grant*. New York, 1935.

Holmes, Oliver Wendell. *Holmes-Pollock Letters*. Edited by M. De Wolfe Howe. Cambridge, Mass., 1946.

* Holt, Henry. *Garrulities of an Octogenarian Editor*. Boston, 1923.

Howe, M. A. DeWolfe. *James Ford Rhodes*. New York, 1929.

———— *Portrait of an Independent: Moorfield Storey 1845-1929*. Boston, 1932.

———— *John Jay Chapman*. Boston, 1937.

———— *Barrett Wendell and His Letters*. Boston, 1924.

Howells, William Dean, *Life in Letters of William Dean Howells*. Edited by Mildred Howells. New York, 1928.

Huxley, Thomas Henry. *Life and Letters of T. H. Huxley*. Edited by Leonard Huxley. New York, 1902.

Jacks, Lawrence P. *Life and Letters of Stopford Brooke*. London, 1917.

James, Henry. *Charles W. Eliot*. Boston, 1930.

James, Henry. *Letters of Henry James*. Edited by Percy Lubbock. London, 1920.

* James, William. *Letters of William James*. Edited by Henry James. Boston, 1920.

° Jameson, J. F. "The American Historical Review 1895-1920," *American Historical Review*, XXVI.

Joyner, Fred. *David Ames Wells*. Cedar Rapids, 1939.

Kellen, William. "Memoir of Winslow Warren, Class of 1858," *Proceedings*, Massachusetts Historical Society, 1930-1932, vol. 64.

Laughlin, James L. "Some Recollections of Henry Adams," *Scribner's Magazine*, vol. 69, May 1921.

* Lodge, Henry Cabot. *Early Memories*. New York, 1913.

Longfellow, Samuel. *Life of Henry Wadsworth Longfellow*. Boston, 1886.

Lowell, James Russell. *Letters of James Russell Lowell*. Edited by Charles Eliot Norton. New York, 1894.

———— *New Letters of James Russell Lowell*. New York, 1932. Edited by M. A. DeWolfe Howe.

Lyell, Sir Charles. *Life, Letters, and Journals of Sir Charles Lyell*. Edited by his sister-in-law, Mrs. Lyell. London, 1881.

Merriam, George S. *Life and Times of Samuel Bowles*. New York, 1885.

Mitchell, Stewart. "Henry Adams and Some of His Students,"

Proceedings, Massachusetts Historical Society, 1936-1941, vol. 66.

* Monroe, James P. *Life of Francis Amasa Walker.* New York, 1932.

Moran, Benjamin. "Diary." Unpublished manuscript in the Library of Congress.

Morison, Samuel E. "Edward Channing," *Proceedings,* Massachusetts Historical Society, 1930-1932, vol. 64.

Morse, John T. *Thomas Sergeant Perry.* Boston, 1929.

———— *Life and Letters of Oliver Wendell Holmes.* Boston, 1869.

———— "Incidents Connected with the American Statesmen Series," *Proceedings,* Massachusetts Historical Society, 1930-1932, vol. 64.

Motley, John Lothrop. *The Correspondence of John Lothrop Motley.* Edited by George W. Curtis. New York, 1889.

Muzzey, David S. *James G. Blaine.* New York, 1934.

Nevins, Allan. *Hamilton Fish: The Inner History of the Grant Administration.* New York, 1936.

* ———— *Abram S. Hewitt.* New York, 1935.

Nicolay, J. G. and J. Hay. *Abraham Lincoln.* New York, 1890.

Norton, Charles Eliot. *Letters of Mrs. Gaskell and Charles Eliot Norton.* London, 1922.

* Ogden, Rollo. *Life and Letters of Edwin Lawrence Godkin.* New York, 1907.

Pearson, Henry G. *Life of John A. Andrew.* Boston, 1904.

* Perry, Bliss. *Life and Letters of Henry L. Higginson.* Boston, 1921.

Perry, Ralph Barton. *The Thought and Character of William James.* Boston, 1935.

Pike, James S. "From the Diaries of a Diplomat," edited by Harold Davis, *New England Quarterly,* 1941, XIV.

Reid, T. Wemyss. *Life, Letters, and Friendships of Richard Monckton Milnes.* New York, 1891.

* Rhodes, James Ford. "Henry Adams, '58," *Harvard Graduates Magazine,* 1917-18, vol. 26.

° ———— [Henry Adams, a tribute]. *Proceedings,* Massachusetts Historical Society, 1917-18, vol. 51.

Robinson, William S. *"Warrington" Pen-Portraits.* Boston, 1877.

Russell, Lady John. *Lady John Russell: A Memoir with Selections from Her Diaries and Correspondence.* Edited by Desmond McCarthy and Agatha Russell. New York, 1911.

Ruykeyser, Muriel. *Willard Gibbs.* New York, 1942.

* St. Gaudens, Augustus. *The Reminiscences of Augustus Saint-Gaudens.* Edited by Homer St. Gaudens. London, 1913.

Schurz, Carl. *Speeches, Correspondence, and Political Papers.* Edited by Frederic Bancroft. New York, 1913.

—— *Intimate Letters of Carl Schurz. 1841-1869.* Madison, 1928.

* Sedgwick, Henry. *Francis Parkman.* Boston, 1904.

Seward, Frederick. *William H. Seward.* New York, 1891.

Smith, Theodore. *The Life and Letters of James Abram Garfield.* New Haven, 1925.

Steiner, Bernhard. *Life of Henry Winter Davis.* Baltimore, 1916.

Stephens, William R. *Life and Letters of Edward Augustus Freeman.* London, 1895.

* Stern, Bernhard J. *Lewis Henry Morgan.* Chicago, 1931.

Swift, Lindsay. "A Course in History at Harvard College in the Seventies," *Proceedings,* Massachusetts Historical Society, 1918-19, vol. 52.

Taylor, Bayard. *Unpublished Letters of Bayard Taylor.* Edited by John R. Schultz. San Marino, Calif., 1937.

* Taylor, Henry Osborn. *Human Values and Verities, Part I.* Privately printed, 1929.

* —— "The Education of Henry Adams," *Atlantic Monthly,* October 1918.

* Thayer, William Roscoe. *The Life of John Hay.* Boston, 1915.

—— *Letters of William Roscoe Thayer.* Edited by Charles Hazen. Boston, 1926.

Thwing, Charles F. *Guides, Philosophers, and Friends.* New York, 1927.

* Tilden, Samuel J. *Letters and Literary Memorials of Samuel J. Tilden.* Edited by John Bigelow. New York, 1908.

* Wade, Mason. *Francis Parkman.* New York, 1942.

Washburn, Charles G. "Memoir of Henry Cabot Lodge," *Proceedings,* Massachusetts Historical Society, 1924-25, vol. 58.

Watterson, Henry. *'Marse Henry' An Autobiography.* New York, 1919.

Welles, Gideon. *Diary of Gideon Welles.* Boston, 1911.

White, J. C. "An Undergraduate's Diary," *Harvard Magazine,* vol. XXI, 423-430, 636-651.

Williams, Charles R. *Life of Rutherford B. Hayes.* Boston, 1914.

Williamson, Harold F. *Edward Atkinson.* Boston, 1934.

Willis, Henry Parker. *Stephen A. Douglas.* Philadelphia, 1910.

Winslow, Warren. "Recollections of Fifty Years," *Proceedings,* Massachusetts Historical Society, 1922-23, vol. 56.

Winsor, Justin. *Charles Deane.* Cambridge, 1891.

—— (ed.) *Bibliographical Contributions,* No. 12. List of the Publications of Harvard University and Its Officers. 1870-1880. Cambridge, 1881.

Woolner, Amy. *Thomas Woolner, R.A. Sculptor and Poet.* London, 1917.

HISTORICAL STUDIES AND GENERAL WORKS

(See footnotes for works read by Henry Adams.)

Adams, Charles Francis. *Address on the Life of Seward.* New York, 1873.

—— *Address to the Phi Beta Kappa Society.* Cambridge, 1873.

—— *An Appeal from the New to the Old Whigs . . . by a Whig of the Old School.* Boston, 1835.

—— *Reflections upon the Present State of the Currency.* Boston, 1837.

Adams, Charles Francis, Jr. *Lee's Centennial.* Boston, 1907.

—— *Studies Military and Diplomatic.* New York, 1911.

Adams, Ephraim D. *Great Britain and the American Civil War.* London, 1925.

Adams, Frank D. *Birth and Development of the Geological Sciences.* Baltimore, 1938.

Adams, Herbert B. *Study of History in American Colleges and Universities.* Washington, 1887.

Adams, John. *Works.* Edited by Charles Francis Adams. Boston, 1850-1856.

Adams, John Quincy. *Letters and Addresses on Freemasonry.* Dayton, 1875.

Agassiz, Louis. *Methods of Study.* Boston, 1863.

—— *Contributions to the Natural History of the United States.* Boston, 1857-1877.

Bacon, Edwin. *Boston—A Guide Book.* Boston, 1928.

Beale, Howard K. "On Rewriting Reconstruction History," *American Historical Review,* 1939-40, vol. 45.

Beard, Charles A. "Teutonic Origin of Representative Government," *American Political Science Review,* February 1932, vol. 26.

Bemis, Samuel F. *Guide to the Diplomatic History of the United States.* Washington, 1935.

Bernard, Montague. *Historical Account of the Neutrality of Great Britain during the Civil War.* London, 1870.

Cargill, Oscar. "The Medievalism of Henry Adams," *Essays and Studies in Honor of Carleton Brown.* New York, 1940.

Channing, Edward. *A History of the United States.* New York, 1920.

Cole, Arthur C. *The Irrepressible Conflict 1850-1865.* New York, 1934.

Coleman, Charles. *The Election of 1868.* New York, 1933.

Conant, James. *History and Traditions of Harvard College.* Cambridge, 1934.

Craven, Avery. *The Coming of the Civil War.* New York, 1942.

Davis, Elmer. *History of the New York Times.* New York, 1921.

Dodd, William E. *Expansion and Conflict.* Boston, 1915.

Elliott, E. N. *Cotton Is King, and Pro-Slavery Arguments.* Augusta, Ga., 1860.

Ford, Henry J. *Representative Government.* New York, 1924.

Foster, W. E. *Literature of Civil Service Reform.* Providence, 1881.

Gabriel, Ralph. *The Course of American Democratic Thought.* New York, 1940.

Glover, Gilbert. *Immediate Pre-Civil War Compromise Efforts.* Nashville, 1934.

Gooch, G. P. *History and Historians of the Nineteenth Century.* London, 1913.

Harvard University. *Catalogue of the Officers and Students.*

——— *Catalogue of the Harvard Chapter of Phi Beta Kappa.* Cambridge, 1912.

——— *Centennial History of the Harvard Law School.* Boston, 1918.

——— *Historical Register of Harvard University.* 1636-1936.

——— *Hasty Pudding Club. Annual Catalogue.*

——— *Orders and Regulations of Harvard University.* July, 1855.

——— *Report of the Class of 1858.* James C. Davis. Boston, 1898.

Hicks, Fredrick, ed., *High Finance in the Sixties.* New Haven, 1929.

Holmes, Pauline. *Tercentenary History of the Boston Public Latin School.* Cambridge, 1935.

Howard, George E. *History of Matrimonial Institutions.* Chicago, 1904.

Josephson, Matthew. *The Robber Barons.* New York, 1934.

Koren, John. *Boston 1822-1922. Story of Its Government.* Boston, 1922.

Kraus, Michael. *A History of American History.* New York, 1937.

Leech, Margaret. *Reveille in Washington.* New York, 1941.

Long, O. W. *Literary Pioneers: Early American Explorers of European Culture.* Cambridge, 1935.

McCabe, James D. *Behind the Scenes in Washington.* New York, 1873.

McClellan, George. *Modern Italy.* Princeton, 1933.

McPherson's Handbook of Politics. Washington, 1872.

Marraro, Howard. *American Opinion on the Unification of Italy.* New York, 1932.

Matthews, Albert. *Harvard Commencement Days*. Cambridge, 1916.

Milton, George F. *The Eve of Conflict*. Boston, 1934.

Mitchell, Wesley. *A History of the Greenbacks*. Chicago, 1903.

Morison, Samuel E. *Three Centuries of Harvard*. Cambridge, 1936.

—— ed., *The Development of Harvard University*. Cambridge, 1930.

—— and Henry S. Commager. *Growth of the American Republic*. New York, 1937.

Mott, Frank L. *American Journalism*. New York, 1941.

—— *A History of American Magazines*. Cambridge, 1938.

—— "One Hundred and Twenty Years," *North American Review*, vol. 240 (June 1935), pp. 144-174.

Mowat, Robert. *The Diplomatic Relations of Great Britain and the United States*. London, 1925.

—— *The States of Europe 1815-1871*. New York, 1932.

Nevins, Allan. *The Evening Post*. New York, 1922.

Owsley, Frank. *King Cotton Diplomacy*. Chicago, 1931.

Palgrave, Sir Francis. *History of Normandy and England*. London, 1851-1864.

Pollak, Gustav. *Fifty Years of American Idealism: The New York Nation 1865-1915*. Boston, 1915.

—— "The 'Nation' and Its Contributors," *Nation*, vol. 100, July 8, 1915.

Pollard, Edward. *The Lost Cause*. New York, 1867.

Quincy, Josiah. *History of Harvard University*. Cambridge, 1840.

Ratner, Sidney. "Was the Supreme Court Packed by President Grant?" *Political Science Quarterly*, vol. 50. September 1935.

Rhodes, James F. *Historical Essays*. New York, 1909.

—— *History of the United States*. New York, 1906.

Rose, L. A. "A Bibliographical Survey of Economic and Political Writings 1865-1900," *American Literature*, vol. 15 (January 1944), pp. 381-400.

Ross, E. D. *The Liberal Republican Movement*. New York, 1919.

Schlesinger, Arthur, Jr. *The Age of Jackson*. Boston, 1945.

Schouler, William. *A History of Massachusetts during the Civil War*. Boston, 1868.

Tansill, Charles. *The United States and Santo Domingo: 1798-1873*. Baltimore, 1938.

Taylor, Alan. *Italian Problem in European Diplomacy*. Manchester, 1934.

Thwaites, Reuben. *Outline of Glacial Geology*. Ann Arbor, 1934.

Trevelyan, George. *Garibaldi and His Thousand*. London, 1910.

Wallace, David. *History of South Carolina*. New York, 1934.

Ware, Edith E. *Political Opinion in Massachusetts during Civil War and Reconstruction.* New York, 1916.

Warren, Charles. *The Supreme Court in United States History.* Boston, 1922.

Whittaker, Edmund. *History of Economic Ideas.* New York, 1940.

Zucker, Morris. *Philosophy of American History.* New York, 1945.

INDEX

DATE DUE

GAYLORD			PRINTED IN U.S.A.